The Cosmospheric Principle

A Systemic Modeling of the Universe

PAUL TALBOT

 v5 ISBN 978-2-9819000-3-6

To Yeshua, Jesus and Isa

FOREWORD

This book deals mainly with cosmology. Although some physical and mathematical aspects are addressed, it is first a conceptual approach, which involves philosophical and historical considerations of the subject. It seeks to bridge the gap between metaphysics and physics, and suggests a challenge to the strictly objective interpretation of contemporary cosmology. To do so, a systemic approach is used. The Universe is thus no longer considered as a simple object, but as a system articulated around a new unifying principle: The Cosmospheric Principle.

In writing this book, I am simply trying to explain how I see the world after several years of research and reflection on the subject. In a field as vast and complex as cosmology, I believe it is useful to add to the scientific rigor, some notions from other disciplines, such as ontology, theology and the history of the philosophy of nature.

Modern physics offers us the image of a universe devoid of meaning and purpose. It would have created itself, out of nowhere, by chance. It would be governed by an increasing entropy which should logically lead to chaos. Instead, it would have generated living and conscious beings, still by chance. Our very existence would be meaningless and the free will we think we exercise would be an illusion! Indeed, the Universe being deterministic, the future would already exist, somewhere in a predetermined space-time.

This book argues that this disconcerting reading of our reality is based on unjustified assumptions.

Many believe that modern cosmology is in crisis. Initial singularity, eternal inflation, multiple universes and worlds, superluminal velocities, exotic dark matter, dark energy, extra dimensions and simulated universes are only examples that illustrate the impasse in which cosmology is now plunged. In spite of several hypotheses that are often unverifiable, we still do not manage to explain simply the cosmological observations and the world that surrounds us.

Two French-written books by astrophysicist, historian and philosopher Thomas Lepeltier could help explain my approach. (La face cachée de l'Univers - Une autre histoire de la cosmologie, 2014) *(The Hidden Face of the Universe - Another History of Cosmology)* exposes the historical and methodological reasons that explain a persistent and growing skepticism towards some elements of the modern cosmological paradigm. A more recent work (L'Univers existe-t-il ?, 2021) *(Does the Universe exist?)* challenges the very relevance of picturing the Universe as an object of strictly scientific study.

It is legitimate to question the limits of the scientific method in this field. The observable universe is by definition unique. We cannot therefore recreate a universe in the laboratory, observe its evolution and repeat the experiment. Moreover, our window of observation on the Cosmos is tiny compared to the dimensions of the observable universe. Relevant observations have been made within the limits of the solar system for only a few decades.

Regarding what is not observable, one can formulate hypotheses, but these do not really belong to science, since they can neither be corroborated nor refuted by observation.

For these reasons, I believe that science alone does not allow us to deal effectively with the cosmological question. In order to progress in my understanding of the Cosmos, I believe it is required to step back and reconsider certain fundamental concepts such as time and space. This ontological questioning naturally led me to the concept of Cosmosphere and the principle which follows from it.

Using all available means, I particularly appreciate the information coming from the free encyclopedia Wikipedia. I would like to take this opportunity to highlight the contribution of volunteers who often know how to make accessible, information that was once reserved for specialists.

I also welcome the work of specialists in the field who bother to popularize content that is often difficult to access, with special mentions for Sir Roger Penrose (The Road to Reality, 2004) and Mr. Luke A. Barnes and Geraint F. Lewis (The Cosmic Revolutionary's Handbook, 2020).

References and quotations in this book do not imply that cited authors share the opinions expressed or the assumptions made.

This English edition is a translation of the French-written book (Talbot, Le principe cosmosphérique, 2021).

INTRODUCTION

In the last century or so, modern physics has begun to question objective reality. On a tiny scale, we encounter a minimum: the quantum of action. Some microscopic phenomena can no longer be described exclusively in terms of material particles or waves. New concepts such as the quantum state, the wave function and the probability of presence allow describing phenomena with great precision, but the correspondence of these concepts with our macroscopic reality remains difficult to grasp.

As a result of recent astrophysical observations, new exotic entities such as dark matter and dark energy have emerged. The existence of these entities relies on the interpretation of observations in the light of current theories and hypotheses, but in fact, very little is known. To account for several phenomena at different scales, this book suggests the existence of a maximum: The Cosmosphere.

We live in a fertile era in cosmology, and new models are continually being proposed. Here are three reasons why a curious mind might be interested in the cosmospheric principle:

- It implies a coherent ontology, and relies on it.
- It explains several phenomena, now poorly understood.
- It shows a high level of compliance with observation.

Although the language used is relatively accessible, some sections require knowledge in science and philosophy. More specifically, some notions of physics, astronomy, ontology and epistemology are used.

Despite its name, the cosmospheric principle is neither geocentric nor anthropocentric; it is not a return of the closed world of antiquity, centered on humanity. Inhabitants of a distant galaxy or another cosmic epoch could very well use this principle to explain the world around them. Seen in this light, the cosmospheric principle has a certain form of universality: it is assumed to be valid anytime and anywhere. References to the history of cosmological thoughts essentially aim to better understand the evolution of the concepts leading to the formulation of this principle.

However, the cosmospheric principle is incompatible with the Copernican principle. It rejects spatial homogeneity and the uniformity of the flow of time. It assigns to the observer a central role in the observation process and proposes quite a different vision of the cosmos. As it implies an alternative paradigm to that of the Big Bang, a fair open-mindedness will facilitate the assimilation of the proposed concepts and principles.

This book is not only about the Universe, but about the way we perceive it too, the prism through which we observe it. What is questioned is not the observations as such, but the interpretation we make of them.

The approach used seeks to overcome some forms of cognitive bias, including the dogma of objective realism. It presupposes the maintenance of reasonable doubt on the alleged obviousness or other received ideas. When required, these will be subjected to the test of reasoning and observation. It is an invitation to a structured reflection on our perception of the world.

The cosmospheric principle is inspired by a metaphysical system introduced by Cardinal Nicholas of Cusa in the Middle Ages. This metaphysics bears the name *«On Learned Ignorance»* and includes a cosmological component based on a panentheistic theology from a Christian perspective. The roots of this theology themselves go back to antiquity. The cosmological system described in this book is compatible with the existence of a personal and transcendent God who has a particularly fruitful relationship with the Universe he himself generated.

I wish you a good trip in this captivating vision of the cosmos that is the cosmospheric universe.

Limits and Expectations

The cosmospheric principle provides a coherent interpretation of time, space, mass and motion. It allows explaining several phenomena on cosmological, macroscopic and quantum scales. However, its explanatory power is limited. For example, the Universe seems to have properties of emergence and self-organization. It has allowed the evolution of complexity and life, up to forms as evolved as the human being and all human achievements. These very important questions are only partially addressed in this book.

The cosmospheric principle alone does not explain all the forces of nature and the particles associated with them. It is not a theory of everything, but simply the result of a conceptual approach that attempts to apply the principles of the systemic approach to the Universe as a whole. Because of the magnitude of the subject matter and the limits of my knowledge, some of the hypotheses formulated can rightly be qualified as speculative. However, I believe that some of them are worth considering, because they are coherent and corroborated by observation.

The cosmospheric system could serve as a framework for those who would like to pursue this path. Systemic cosmology could interest several categories of specialists including physicists, philosophers, historians, mathematicians and theologians.

Content and Text Structure

The writing of the text aims at a progressive introduction to the concept of Cosmosphere, its properties and the principle which results from it. The argumentation is sometimes philosophical, historical, physical or mathematical. The final objective is to illustrate the cosmospheric interpretation of phenomena, so targeted observations, supporting this interpretation, are included in the text.

It contains a cosmological system with compatible hypotheses, concepts and relationships, as well as a model of physical quantities based on this system. The cosmospheric hypothesis implies that the past of the Universe would differ significantly from what conventional theories suggest. The same applies to the future of the Universe in the very long term. This book includes an exploratory section that attempts to extrapolate some aspects of this cosmospheric past and future.

Notation of Physical Quantities and Units

The notation used for physical quantities and units is generally in accordance with the "International System of Units" (SI). The values used for physical constants are those of CODATA 2018, and are listed in Appendix A.

Regarding astronomical quantities, this work follows the recommendations of the International Astronomical Union (IAU). The unit of time "Julian year" (noted a) is defined as 365.25 days. The astronomical unit (noted au) is roughly the distance between the earth and the sun. Its value is exactly 149 597 870.7 km.

The light-year (noted ly) is the distance traveled by light in vacuum during a Julian year. This distance is about 63 241 au, or 9 460 730 472 580.8 km. The parsec (noted pc) is sometimes used and corresponds to a distance of 206 264.8 au, that is to say approximately 3.261 6 ly.

In accordance with the SI, and to avoid ambiguity, the term "billion" is not used in this book. The prefix G (for giga) is used instead.

The Constructivist Perspective

We are all, consciously or not, influenced by our beliefs about the world around us. Many believe, for example, in the existence of an objective reality, independent of any observer. Physical laws and mathematical equations would allow describing this reality. This perspective is called physicalism, a modern version of materialism.

Realistic cosmologies formulate assertions about the very nature of the Universe: a geometric object, a mathematical equation, a quantum computer, a hologram or something similar. The Universe has long been regarded as a clockwork mechanism whose cogs fit together perfectly. In fact, the Universe seems to share some properties of the above examples. No one denies, for example, that stars naturally constitute a clock, a calendar or a compass. It is rather the restrictive interpretation of these concepts that is problematic. If we limit our perception of the Universe to one of these notions, then this perception is necessarily incomplete.

The philosophical point of view adopted in this book is different. While acknowledging the existence of a reality which the observer is part of, it considers that we do not have direct access to this reality (the thing-in-itself according to Kant). However, our senses and the concepts we forge would allow us to apprehend it, at least partially through the phenomenon. This explains why we can conceive the Universe in diverse ways. This philosophical perspective is constructivism. If science and technology have allowed broadening the spectrum of our perceptions, and refining observations, the evolution of concepts is a distinct field, which largely motivates this book.

Spinoza is credited with the expression: *«The concept of dog does not bark».* The objective is therefore not to describe reality per se, but to represent (or conceive) it, to be able to account for phenomena coherently.

In his book *«De la vérité dans les sciences» (On Truth in Sciences*), the astrophysicist and philosopher Aurélien Barrau writes:

«Truth is obviously a necessary guide that should not be denied in any way, but it is itself dependent on a partially constructed system of thought...»

«Science is a way of making a world, among other possibilities. It is coherent and convincing, it is beautiful and rational, but it does not necessarily have a privileged relationship with truth or with reality per se.» (Barrau, 2016)

For this reason, the text is often written in the conditional, emphasizing that what is formulated depends on the point of view adopted (the cosmospheric point of view), and does not refer to an objective, unique and absolute reality. However, the conditional is sometimes omitted so as not to make the text unwieldy. Overall, we can consider that the cosmospheric system stems from a Kantian epistemology.

The Systemic Approach and Modeling

Systemic is a global approach that considers the system as a whole, whose dynamic is ensured by a continuous interaction of its components. That is what is meant by saying that the whole cannot be reduced to the sum of its parts. For example, the human body is not limited to the assembly of a trunk, a head, arms and legs. Nor does the systemic reduce it to an assembly of organs, cells, molecules or atoms. Rather, it is assimilated to a complex system whose harmonious functioning requires the continuous interaction of several interdependent subsystems: nervous, blood, muscular, digestive... Systemic thus aims to go beyond the limits of the analytical method and other forms of reductionism. The objective is modeling, i.e. the representation of a complex reality using a simplified model. The philosopher and scientist Alfred Korzybski illustrates the non-realistic nature of modeling as follows: *«A map is not the territory.»* Modeling defines some concepts and makes them interact to represent the coherent and efficient functioning of a system. Ideally, this representation will be simple and aesthetic.

The Cosmological Puzzle

Cosmological modeling can be compared to assembling a puzzle with several pieces missing. We also have the impression that some pieces are mismatched or surplus. The systemic approach allows us to form a global image of the puzzle, the one usually found on the lid of the box. Without this global image, it is difficult to correctly arrange what we have to form a coherent whole. On the other hand, if we have such an image, we can try to imagine the missing pieces that would allow us to integrate harmoniously those that seem to be mismatched.

For example, the proposed system must be able to reconcile relativistic observations with those of quantum physics. In other words, it must be possible to explain why the speed of light constitutes a physical maximum, whereas the quantum of action corresponds to an indivisible minimum. Conventional theories pose these axioms based on observation but do not really provide an explanation. This is an inherent limit of the scientific method. For this laborious task, I believe it is useful to use all the tools and resources available. Imagination can be used in a structured approach. The evolution of cosmological thinking can also be called upon. Modern thinkers as well as those of antiquity and the Middle Ages may have something to teach us about this global picture and these missing puzzle pieces.

The System Design Process

Cosmological modeling can also be assimilated to an optimization process. The aim is to maximize the coherence, efficiency, simplicity and aesthetics of the system with a minimum of hypotheses, concepts and principles that are compatible with each other. This is a generalization of the principle of simplicity or Occam's razor, which suggests that the explanation requiring the fewest hypotheses is the most likely. What a cosmological system tries to explain is the set of all observable phenomena, hence the need to restrict the number of hypotheses, principles and concepts.

The process used is iterative, and calls for imagination and reasoning in a synergistic mode. Concepts and their interactions are iteratively evaluated by different types of reasoning (deduction, induction, abduction and intuition) according to the objective: the coherent, efficient, simple and aesthetic functioning of the system. Depending on the results of this evaluation, these concepts and their interactions are either retained, discarded, modified or reorganized to improve the system. If necessary, new concepts or principles will be introduced to better account for phenomena than others, which then become superfluous.

The working hypothesis is that by tightening the conceptual vise, the resulting model will possess the desired characteristics. Theoretically, this process is endless, and the system can always be improved.

The writing of the text seeks to produce a structured sequence of reasoning, but the approach itself is not linear. The process must also be adapted to the domain for which the system is intended. For a cosmological system, particular attention must be paid to the conformity of the model to observation. Part of the work consists of establishing cross-references between different observations, and trying to explain them using some concepts, principles or theories already proposed. The starting point is the assimilation of generally accepted concepts, and here I can only describe the progress of my work.

For some, this approach may seem unnecessary, but several factors need to be considered. On the one hand, system modeling and architecture techniques have greatly evolved. On the other hand, there is now a vast array of increasingly varied and accurate observations. Finally, we can count on the increased and rapid availability of relevant information. Thanks to the Internet, access to this information is often only a mouse click away.

The systemic processes of architecture, design and modeling are powerful tools that allow us to move efficiently towards a rational and global understanding of the Cosmos, by providing a coherent and unified representation of it.

Main Types of Reasoning Used

The main types of reasoning (or inferences) used can be summarized using the following examples, inspired by the philosopher Charles Sanders Pierce.

On a table, there are several beans and a bag already opened:

Deduction:

- *Premise:* This bag contains white beans.
- *Premise:* These beans come from this bag.
- *Conclusion:* These beans are white.

Induction:

- *Premise:* These beans come from this bag.
- *Premise:* These beans are white.
- *Probable conclusion:* This bag contains white beans.

Abduction:

- *Premise:* This bag contains white beans.
- *Premise:* These white beans come from a bag.
- *Probable conclusion:* These white beans come from this bag.

There are several forms of abduction, including one that stands out from the others and which I would describe as intuition.

Intuition:

- *Premise:* This bag contains something.
- *Premise:* These white beans come from somewhere.
- *Probable conclusion (1):* This bag contains white beans.
- *Probable conclusion (2):* These white beans come from this bag.

What distinguishes intuition from other types of inferences is that it leads to several probable and highly correlated conclusions. In the example, one seeks to know both the contents of the bag and the origin of the beans. It is the combination of these two questions that leads to the conclusion.

If we assume (1), then we can conclude (2) by abduction. Conversely, if we assume (2), we can conclude (1) by induction. Since each of the conclusions assumes the other by valid reasoning, this form of intuition can be considered a valid type of inference.

The strong correlation between the two conclusions stems from an implicit, but unstated, assumption. In the example, it is assumed that a bag of white beans would have been handled near the table, and some of its contents would have been poured there. The same bag would then have been left on the table. This unique hypothesis then allows explaining several different observations, which are now linked together. It explains not only the origin of the beans and the contents of the bag but also their presence on the table.

Any cosmological approach implies a similar questioning about the content of the Universe and the origin of its components. It is therefore normal that the cosmic modeling process makes use of this intuitive form of reasoning.

The Evolution of Concepts

Some concepts are useful for a while and then become superfluous afterward. For example, until the beginning of the 20th century, the concept of ether was used to explain the propagation of light waves in vacuum and other phenomena acting at a distance. It is practically no longer used today. The case of the ether is not unique. We can think of phlogiston, the fluid supposedly responsible for combustion, or the epicycles that have allowed the maintenance of geocentrism for several centuries. Perhaps by reconsidering objective reality on a large scale, we could free ourselves from some rather ethereal concepts.

Conversely, certain concepts, hypotheses or theories are discarded for a certain duration, and are later rehabilitated. For example, heliocentrism was clearly postulated by Aristarchus of Samos, 280 BC. His disciple Archimedes refutes this hypothesis, based on then-available observations, and supports Aristotle's geocentrism which will prevail for more than 18 centuries. About 400 years BC, Democritus imagined a universe made up of atoms and vacuum. This atomism will remain marginal and speculative until the beginning of the 20th century.

In a conceptual approach, it may be useful to revisit certain concepts put on a siding. The results of my work suggest reconsidering the concept of Cosmosphere, little known and forgotten for more than three centuries.

THE LEARNED IGNORANCE AND PHYSICAL QUANTITIES

Cartesian rationalism, whose influence is omnipresent, is put on hold here. We are rather interested in the doctrine of Nicholas of Cusa: The Learned Ignorance. The concept of Cosmosphere is present, and the cosmospheric principle is greatly inspired by the Learned Ignorance. We can consider the cosmospheric system as an extension of the cusanian cosmology, which would consider the achievements of modern science. Therefore the underlying ontology and physical interpretation of phenomena differ from conventional theories.

The Legacy of the Learned Ignorance

In 1440, Galileo had not yet pointed his telescope towards the stars, Copernicus had not published his work and Christopher Columbus had not yet set foot on the soil of the Americas. We were in the Middle Ages and none of these characters had yet been born. It is in this context that Cardinal Nicholas Krebs (of Cusa) wrote a book entitled *«On Learned Ignorance».* Socrates' philosophy is here erected as a metaphysical system. As for many medieval works, the reading of the Learned Ignorance can be difficult; the text sometimes seems obscure, even erroneous. Faith, metaphysics, mathematics, philosophy of nature, and some hermetic beliefs are intertwined in a Rhenish mysticism, inherited from a Christian Neoplatonism. It is therefore not a scientific work, but it is considered as a precursor work in many respects.

Nicholas of Cusa breaks with Aristotle's scholasticism and physics. He decenters the earth from the world and sets it in motion. He shatters the sphere of the fixed stars, which Copernicus will not dare to do a century later:

«The earth, therefore, which cannot be the center, is, however, not devoid of all movement... Thus, just as the earth is not the center of the world, so the sphere of the fixed stars is not its circumference.» (Cues, Docte ignorance, 2008, p. 161.)

Almost two centuries before Galileo, Nicholas of Cusa postulates the relativity of movement:

«For us, from now on, it is clear that the Earth is really moving, even if it doesn't appear so. Indeed, we only apprehend movement in relation to a fixed point. For example, if someone, being in a boat in the middle of the water, did not know that the water is flowing and did not see the shore, how would he apprehend that he is moving ? Thus, it will always seem to the person on Earth, in the Sun or another star, that he is stationary at the center and that all other things are in motion.» (Cues, Docte ignorance, 2008, p. 165)

Nicholas of Cusa does not clearly identify the rotational and translational movements of the earth around the sun. However, he puts the earth in orbit like the other stars, and states that these orbits are not perfectly circular, which Johannes Kepler will confirm in the 17th century:

«Even more, neither the Sun, nor the Moon, nor the Earth, nor any sphere, even if it appears otherwise to our eyes, can in their movement describe a true circle, because they do not move around a fixed point.» (Cues, Docte ignorance, 2008, p. 163)

«The figure of the Earth is therefore noble and spherical, and its movement is circular, but it could be more perfect.» (Cues, Docte ignorance, 2008, p. 166)

He affirms the existence of extraterrestrial beings:

«Therefore, the inhabitants of the other stars, whoever they are, have no common measure with the inhabitants of this world ... Therefore, as this whole region is unknown to us, these inhabitants remain totally unknown.» (Cues, Docte ignorance, 2008, p. 169)

Maybe even intelligent:

«Or do the forms of any region lie in a higher form, namely the intellectual form, and through this, reach the end which is the purpose of the world ?» (Cues, Docte ignorance, 2008, p. 171)

Since Aristotle, it was believed that the Cosmos was divided into two zones: a sublunar zone, corruptible and governed by terrestrial physics; and a supralunar zone, incorruptible and governed by celestial laws. Nicholas of Cusa clearly freed himself from the dogma of the incorruptibility of the supra-lunar world:

«Indeed, the universe being unique and all the particular stars exerting on each other, proportionate influxes, it will not be possible for us to maintain with certainty that something is totally corruptible or is corruptible according to another mode of being.» (Cues, Docte ignorance, 2008, p. 170)

He had understood that the sun is only one star among others, and the earth, one planet, among others. For him, all these stars are in motion, influence each other and describe more or less circular orbits. In addition, these stars are said to harbor different forms of extraterrestrial life. One wonders why and how a man of faith like Nicholas of Cusa could have formulated such hypotheses in a highly dogmatic context devoid of convincing observations. This is undoubtedly due to his great erudition and the privileged information to which he had access, but perhaps also to the very doctrine he taught:

«The ancients have not reached all we have just said because they missed the learned ignorance.» (Cues, Docte ignorance, 2008, p. 165)

The Learned ignorance invites us to become aware of the limits of our knowledge. This awareness facilitates the assimilation of certain concepts and principles, which one would tend to reject a priori, because of their apparent strangeness. This perception often results from the incompatibility of these concepts and principles with others that we take for granted. In other words, our very understanding of phenomena is limited by our purported knowledge. By taking comfort in his certainties, the human being finds himself trapped by his own beliefs. This explains, for example, why an erroneous hypothesis such as geocentrism has lasted so long. Modern psychology today identifies this condition as a cognitive bias: the belief bias that aims precisely at avoiding cognitive dissonance.

It is a fact that science and technology have transformed our island of human knowledge into a supercontinent of erudition. However, if the ocean of ignorance in which this knowledge is bathed is immense, then our limited knowledge is nothing in comparison. As one rises higher and higher above this Pangea of knowledge, it becomes again an island, an islet, and finally an invisible and quite negligible point. Human omniscience is nothing more than a chimera, and like Socrates, we can know that we do not know.

The Maximum and the Minimum of the Learned Ignorance

The Learned Ignorance introduces two fundamental concepts from which the cosmospheric system draws its inspiration:

- The maximum is what cannot be larger.
- The minimum is what cannot be smaller.

These limits would be at the same time physical, absolute and transcendental:

«Now everything that is measurable falls between the maximum and the minimum.» (Cues, Docte ignorance, 2008, p. 78)

«The Maximum and the Minimum, as understood in this work, are transcendental terms, of absolute meaning, so that they embrace all things in their absolute simplicity, beyond any contradiction in a quantity of mass or force.» (Cues, Docte ignorance, 2008, p. 47)

These cusanian Maximum and Minimum coincide with each other and with unity. They transcend any form of opposition, affirmation or negation. Contrary to other values, they are no longer governed by Aristotle's principle of non-contradiction, but rather by that of *«coincidence of opposites»* which constitutes the essence of the cusanian thought. On his WEB site, the philosopher Jean-Marie Nicolle, a specialist in this field, summarizes the purpose of this principle as follows:

«Coincidence of opposites has the cognitive function of finding the link between the diversity of the reality and the unity of the superior principle, in a bottom-up resolving approach.» (Nicolle, s.d.)

This work suggests that Nicholas of Cusa was right, and that he anticipated the existence of physical maximum and minimum. These entities would indeed be governed by a certain form of coincidence of opposites, which yield them a special status.

The quantum of action (noted h) seems to match the concept of physical minimum. Strangely, this minimal entity seems to induce some paradoxical properties of quantum systems. For example, the quantum principle of superposition tells us that the state of a quantum system simultaneously has several values, only one of which is actualized during observation. To illustrate the strangeness of this principle, Erwin Schrödinger proposes the image of a cat that would be both dead and alive at the same time! Quantum superposition can be interpreted as a form of coincidence of opposites (or mutually exclusives) values.

Moreover, we observe that in any isolated physical system, the quantity associated with this quantum minimum (Action) is governed by a minimum principle: the principle of least action. This means that no matter what happens, the number of resulting quanta is always minimal, which is intriguing, to say the least.

In the same vein, Werner Heisenberg, the instigator of the uncertainty principle, wrote in the last century:

«I consider the ambition of overcoming opposites, including also a synthesis embracing both rational understanding and the mystical experience of unity, to be the mythos of our present-day and age.»

If, while reading this book, you feel that certain statements seem to diverge, I suggest that you continue up to the section *«Beyond the Opposites»* which provides the required explanations.

Concerning the Maximum, Nicholas of Cusa explicitly associates it with the Maximum sphere (the Cosmosphere), a unique entity that encompasses the whole cosmos. This maximum would itself be the principle and the measure of the Universe and all that it contains.

Physical Quantities and Their Limits

Inspired by the cusanian Minimum and Maximum, it is appropriate here to introduce the concepts of physical quantity, absolute minimum and absolute maximum. These concepts are essential to the understanding of this book.

A physical quantity would be a quantity that has a physical dimension. For example, Time, Length, Area, Speed, Mass or Force would be physical quantities. However, an angle, an index or a numerical quantity such as a mole does not have a physical dimension. In this book, the concept of physical quantity is restricted. It excludes dimensionless quantities. The basic hypothesis is that the field of physics should be limited to phenomena that are in principle observable, i.e. those that could theoretically be measured. Agreeing with the Learned Ignorance, and to preserve physical quantities from the inconsistencies inherent to infinite values, these would be limited minimally and maximally:

- The absolute maximum of a physical quantity is the maximum and finite value that this quantity can possess. For example, Speed is a maximally limited quantity, and its absolute maximum is the speed of light in vacuum (noted c).
- The absolute minimum of a physical quantity is the minimum non-zero value that this quantity can possess. For example, Action is a minimally limited quantity, and its absolute minimum is the quantum of action (noted h).

The absolute minimum must be distinguished from a null value. For example, we know that Temperature cannot be below -273.15 °C or zero Kelvin (0 K). This value is not the absolute minimum of the quantity Temperature, but rather corresponds to a total absence of heat, which is physically out of reach. There would therefore be a minimum temperature but above 0 K.

It seems to be assumed that the above examples are exceptions, and that physical quantities are generally unlimited. This belief would result in part from the fact that physical limits are either very large or very small. It is relatively recently that some of these have come to light.

The generic term that refers to either the minimum or maximum is «extremum.»

The Ontological Argument

The ontological argument posits that to be in principle observable and measurable, a physical quantity must exist and be finite. Since infinity is by definition immeasurable, this concept would not fall within the scope of physics. Physical quantities would thus be bound by a maximum value. For example, the age of the universe would be finite because we cannot measure an infinite duration.

Likewise, it is risky to assign a physical property to an entity without one. When we turn off the tap, can we consider that the water flow becomes null? We can always look at things from different points of view, as this is the nature of the design process. What the ontological argument is trying to say is that asserting the physical existence of a water flow where there is none may not be the most enlightening viewpoint.

The Argument of the Inverse Quantity

As defined, the physical quantities are invertible. To obtain the inverse of a physical quantity, we simply change the sign (±) of the dimensional exponents that characterize it. For example, in the dimensional system MLT (M: Mass, L: Length and T: Time) the dimension of speed is LT^{-1} (or L/T). The dimension of the inverse speed is therefore $L^{-1}T$ (or T/L).

If a physical quantity is limited by a maximum value, then the inverse quantity is minimally limited by the inverse of this maximum and vice versa. For example, the inverse speed is limited minimally by the inverse of the speed of light ($c^{-1} \approx 3.335\,641$ ns/m). This means that it is impossible to travel one meter in less than 3.335 641 ns. Maximum speed and minimum inverse speed are two ways of expressing the same physical limitation.

Since this inverse speed cannot be infinite, it would itself be maximally limited. The inverse of this maximum would thus constitute the minimum limit of any speed. Applying this reasoning to the other physical quantities, we conclude that they are all limited minimally and maximally.

The Discontinuity Hypothesis

An additional hypothesis is that physical quantities are discontinuous. Indeed, the evolution of such a quantity is characterized by changes of state. For a change of state to be observed or measured, there must be a finite difference between the initial state and the final state. If a physical quantity could take an infinitely small value, then the inverse quantity should be infinitely large, which is excluded. Likewise, if a physical quantity could take a null value, then the inverse quantity should be infinite, which is also excluded. To take the example of the closed tap, if we assume the physical existence of a water flow of 0 L/s, how could we express the inverse of this flow in s/L ?

The Conceptual Convergence

The discontinuous hypothesis, the ontological argument and that of the inverse quantity impose strong constraints on the concept of physical quantity. The rest of the text shows how this tightening of the conceptual vise naturally converges towards the notion of Cosmosphere, whose properties lead just as naturally to the cosmospheric principle.

The Non-Additivity of Speeds

The non-additivity of speeds (or velocities) is a physical property that has been demonstrated and verified experimentally. Indeed, if speeds were additive, then the sum of certain values could exceed the maximum of this quantity (c).

To facilitate the reading of the text, I will call the relativistic composition of velocities *«adjunction»* and I will note this operation $\tilde{+}$: *«almost plus, or adjunct to»*.

If a and b are two speeds expressed in maximum units ($c = 1$), then the equivalence between adjunction and addition is:

$a \tilde{+} b \Leftrightarrow (a+b) / (1+ab)$

For example:
$0.5 \tilde{+} 0.5 \Leftrightarrow (0.5 + 0.5) / (1 + 0.5 \cdot 0.5) = 1/1.25 = 0.8$
and not 1, as suggested by the addition. This equivalence results from the application of the Lorentz transformations, which are used by special relativity. If different units are used, the equivalence between adjunction and addition remains, but involves the maximum speed (c):

$a \tilde{+} b \Leftrightarrow (a+b) / (1+ab/c^2)$

When the velocities involved are much smaller than that of light, the result of the adjunction ($\tilde{+}$) is very close to that of the addition (+). In fact, the result of the adjunction is always less than or equal to that of the addition, as well as to the value of the maximum speed (c).

Are Physical Quantities Additive?

With the exception of Speed, it is generally assumed that other physical quantities, such as Length and Time, are additive. This is indeed what is observed at the human scale, but how to know if this assumption remains valid at the cosmological scale?

According to the approach used, the additivity of physical quantities cannot be taken for granted as scientific knowledge. Rather, it is a widely held belief that must be subjected to the test of reasoning and observation. The fact that we can add up apples or sheep does not imply that we can do the same with meters or seconds.

As we observe that at least one derived quantity (Speed) is not additive, it is not justified to assert a priori that Length and Time are. To consider these quantities as additive is to suppose that they are not maximally limited. Now, from a physical as well as a metaphysical point of view, the infinitude of a duration or a length is just as problematic as that of a speed.

Through reasoning and observation, we can show that physical quantities are not additive. The adjunction ($\tilde{+}$) previously described would apply to several other physical quantities, including Length and Time.

The Non-Additivity of Length and Time

Surprisingly, it can be shown that Length and Time are not additive.

If lengths were additive, then physically, one should observe:

- (1 m + 1 m) = 2 m

By traveling these distances in one second, one should observe:

- (1 m + 1 m) / 1 s = 2 m / 1 s

The division being distributive with respect to the addition, one should observe:

- (1 m / 1 s) + (1 m / 1 s) = 2 m / 1 s, which can be written:
- 1 m/s + 1 m/s = 2 m/s

Now, this is not what we observe. The latter equation is physically inaccurate because it does not use the correct law of composition ($\tilde{+}$). The difference is tiny, but according to the above, we calculate that 1 m/s $\tilde{+}$ 1 m/s = (1+1) / (1+1/c^2) m/s ≈ 1.999 999 999 999 999 978 m/s.

Similarly, if we assume that Time is additive, then physically, we should observe:

- (1 s + 1 s) = 2 s

Accelerating by 1 m/s^2 during these durations, one should observe:

- 1 m/s^2 · (1 s + 1 s) = 1 m/s^2 · 2 s

The multiplication being distributive with respect to the addition, we should observe:

- (1 m/s^2 · 1 s) + (1 m/s^2 · 1 s) = 1 m/s^2 · 2 s

This corresponds to the addition of velocities which is physically inaccurate as previously described:

- 1 m/s + 1 m/s = 2 m/s

These equations show that the additivity of Length or Time implies that of Speed. Since we observe that Speed is not additive, Length and Time would not be additive either. This is called a demonstration by reduction to the absurd.

One might think that the distribution of division or multiplication should not be allowed, because it transforms the addition of lengths or durations into the addition of speeds. This prohibition would automatically imply that Length and Time are not additive. In mathematics, the product of (a) by the sum $(b + c)$ is always equal to the sum of the products $(ab + ac)$:

$$\boldsymbol{a \cdot (b + c) = ab + ac}$$

This is a fundamental property of addition. If this property is not verified, it means that the composition of the involved quantities is not a true addition, even if the result is close to it. This is exactly what is observed in the case of Length and Time.

The explanation for this paradox is that just like speeds, lengths and durations are adjunctive, not additive; these quantities would be limited by maximum values. These limits being very large on a human scale, the results of adjunction for these quantities would be very close to those of the addition.

The conclusion is that all the above equations are physically inaccurate. Time, Length, Area, Volume, and several other derived quantities would not be additive, but rather adjunctive and maximally limited.

This conclusion shakes up preconceived ideas. Minimally and maximally limited quantities would not be the exception, but the norm, as Nicholas of Cusa wrote in 1440! The concept of Cosmosphere (or maximum sphere) follows naturally from this limited extension of time and space, for any observer.

Returning to the previous demonstrations, physically, we would rather observe the following inequalities:

- $(1\ \mathrm{m} \tilde{+} 1\ \mathrm{m}) < 2\ \mathrm{m}$
- $(1\ \mathrm{m} \tilde{+} 1\ \mathrm{m}) / 1\ \mathrm{s} < 2\ \mathrm{m} / 1\ \mathrm{s}$
- $(1\ \mathrm{m} / 1\ \mathrm{s}) \tilde{+} (1\ \mathrm{m} / 1\ \mathrm{s}) < 2\ \mathrm{m} / 1\ \mathrm{s}$
- $1\ \mathrm{m/s} \tilde{+} 1\ \mathrm{m/s} < 2\ \mathrm{m/s}$

- $(1\ \mathrm{s} \tilde{+} 1\ \mathrm{s}) < 2\ \mathrm{s}$
- $1\ \mathrm{m/s^2} \cdot (1\ \mathrm{s} \tilde{+} 1\ \mathrm{s}) < 1\ \mathrm{m/s^2} \cdot 2\ \mathrm{s}$
- $(1\ \mathrm{m/s^2} \cdot 1\ \mathrm{s}) \tilde{+} (1\ \mathrm{m/s^2} \cdot 1\ \mathrm{s}) < 1\ \mathrm{m/s^2} \cdot 2\ \mathrm{s}$
- $1\ \mathrm{m/s} \tilde{+} 1\ \mathrm{m/s} < 2\ \mathrm{m/s}$

We assume here that the adjunction ($\tilde{+}$) always involves the maximum value of the quantity concerned. For now, we assume that the maximum length and maximum duration are very large in relation to the chosen units.

The Objection of Inertial Reference Frames

The objection sometimes raised concerning the previous demonstrations is the use of two inertial reference frames to perform the measurements: one for the moving observer and a second one for a fixed observer. According to the latter, lengths and durations would be somehow additive!

The concept of inertial reference frame is a Galilean idealization which assumes that no force acts on the observer and that no acceleration is involved. For this observer, time would flow uniformly and would be by definition additive. Similarly, space would be homogeneous, isotropic and additive.

According to the cosmospheric system, this idealization does not exist in nature, because it goes against the cosmospheric principle. Instead, we should use the concept of "Cosmospheric Reference Frame" which will be treated in the section "Cosmospheric Interpretation of Motion".

According to the rules of logic, the use of the concept of inertial reference frame to prove the additivity of Length and Time is not allowed, because it implies a circular reasoning. One assumes a priori the existence of an inertial reference frame where Length and Time would be additive to prove the additivity of Length and Time.

QUANTIZED QUANTITIES AND INVERSE QUANTITIES

To formulate a coherent set of hypotheses about physical quantities, the conceptual vise needs to be tightened further. We observe that the minimum value of the quantity Action is a quantum, that is, the atom or the indivisible value that constitutes the increment of this discontinuous quantity. According to the above, like other quantities, Action should be maximally limited. We deduce that the quantum of action and all values of this quantity should be adjunctive, not additive.

If any physical quantity has an absolute minimum, one can wonder if this minimum also constitutes its quantum. This last hypothesis must be rejected because it is inconsistent. The observation of certain inverse quantities refutes the hypothesis of a minimum quantum for these quantities. For example, the absolute minimum of the inverse speed (c^{-1}) is not the quantum of the inverse speed. A decelerating particle does not suddenly change from the inverse speed of $1/c$ to about $2/c$, then to about $3/c$ and so on. Analysis of the distribution of quantized and adjunct values of a quantity shows that those of the inverse quantity are not.

To maintain the consistency of the model, the physical quantities must be divided into two distinct categories: quantized quantities and inverse quantities.

Quantized quantities

Quantized quantities would have a minimum and adjunctive quantum. For example, like the quantity Action, the quantities Time, Length and Speed would each have a minimum and adjunctive quantum.

Mathematically, this model is simple and aesthetic. To obtain the list of values of a quantized quantity (noted Q), we start with the minimum value, the quantum noted q ($Q_1 = q$); for the following values (index $n > 1$), we iteratively apply the adjunction ($\tilde{+}$) of an additional quantum to the previous value: $Q_n = Q_{n-1} \tilde{+} q$.

The result is a discontinuous, quantized, and minimally and maximally limited quantity. Due to the nature of the adjunction, the set of values of a quantized quantity would be infinite, even if the measurement range is finite. For example, by constantly supplying energy to an electron, its speed continuously approaches that of light. However, the electron never reaches this maximum speed because of its mass.

Figure 1 illustrates the distribution of the first 25 values of a quantized but fictitious quantity (Q). The quantum (q) of this quantity would be disproportionately large compared to the maximum (1). This technique allows illustrating simultaneously the quantum and relativistic properties of quantized quantities. For readability purposes, the data illustrated on this graph are listed in Appendix F.

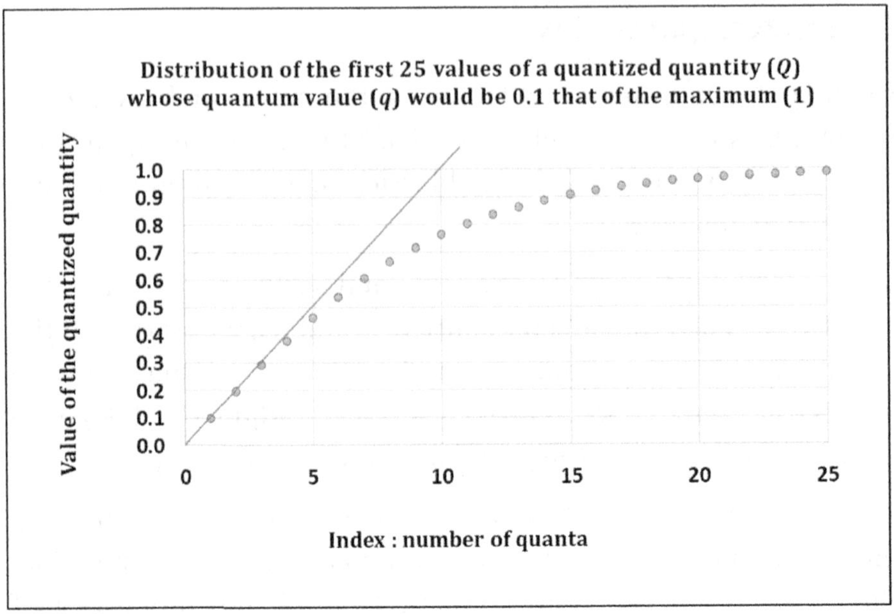

Figure 1 - Fictitious example of a quantized quantity

In this figure, we see that the potential number of quanta is unlimited, although the quantity itself is limited maximally. A quantized quantity, such as length, could not take any positive real value up to infinity (continuous line on the graph). It could only take quantized and adjunct values (points on the graph), just like the quantity action. Moreover, because it is adjunctive, this quantity would be limited maximally, just like the quantity speed.

Inverse Quantities

Inverse quantities would only be the inverse image of quantized quantities. Like quantized quantities, inverse quantities would be discontinuous, and minimally and maximally limited. On the other hand, since they no longer have an adjunctive quantum, the values of these inverse quantities would themselves be non-adjunctive. One way to determine whether a physical quantity is quantized or inverse is to check whether its values are adjunctive or not.

For example, a fuel consumption expressed in liters (L) per 100 km would be adjunctive. Suppose two vehicles in succession consume 10 L/100 km and 20 L/100 km, respectively. The combined fuel consumption of the two vehicles would be 10 L/100 km $\tilde{+}$ 20 L/100 km ≈ 30 L/100 km. I call this quantity quantized and adjunctive, i.e. its physical minimum would be the adjunctive quantum.

On the other hand, fuel consumption expressed in miles per gallon (mpg) would not be adjunctive. Let's assume two vehicles following each other with fuel consumption of 10 mpg and 20 mpg respectively. The combined fuel consumption of the two vehicles is not 30 mpg. Therefore, it would be an inverse, non-quantized and non-adjunctive quantity. To combine values of an inverse quantity, these must first be converted to an adjunctive quantity, such as gallons per mile (gpm), which for the example gives 0.1 gpm $\tilde{+}$ 0.05 gpm ≈ 0.15 gpm. Inverting the result of this adjunction gives the combined consumption in the original unit, which is approximately 6.667 mpg.

HYPOTHESES ABOUT PHYSICAL QUANTITIES

To conclude, I make the following hypotheses about physical quantities:

- Any physical quantity is finite and not null. It is limited both minimally and maximally.
- All physical quantities are discontinuous. The set of values of this quantity is infinite, although the measurement range is finite.
- A quantized quantity has an adjunctive quantum whose value is the minimum of this quantity.
- All values of a quantized quantity are adjunctive according to the law of composition $\tilde{+}$: $a \tilde{+} b$ $(a+b) / (1+ab)$, the unit matching the maximum of this quantity.
- Any quantized quantity can be inverted. Although discontinuous and limited, this inverse quantity is neither quantized nor adjunctive.

The remaining part of the text suggests the existence of an additional, but counterintuitive, hypothesis:

- The extremum values of physical quantities can be associated with a form of spatiotemporal invariance.

THE OBSERVATIONAL SUBJECTIVITY

To interpret the above rationally, it is sufficient to free oneself from the dogma of objective realism. To put it simply, objective realism assumes that reality presents itself to us as it is. This assumption, unfortunately widespread, is contrary to observation.

To illustrate observational subjectivity, the physicist and philosopher Bernard d'Espagnat used the example of the rainbow, which exists independently of any observer, but which nevertheless has properties that depend on him. The cosmospheric principle implies such observational subjectivity of phenomena.

According to the theory of special relativity, time and space are distorted and adjusted so that each observer measures the same value of the maximum speed (c). In a way, they each have their own clock and their own ruler. This implies that the measurement of time and space is not strictly objective, but partly subjective.

The term *«subjective»* here means *«pertaining to the observer, the observing subject».* It can be a person, an animal, a probe or any measuring instrument. It refers to the observational context, not the observer's thoughts, feelings or opinions.

According to the hypotheses previously formulated, we can deduce that any observer must measure the same values for the maximum duration and the maximum length. The invariance of c would naturally emerge from those of the maximum duration and the maximum length.

The Projected Reality

The idea that what we perceive of the world is a projection of reality, and not reality itself, dates to antiquity. In book VII of *«The Republic»,* Plato presents his theory of ideas using the allegory of the cave. Men chained and immobilized in a cave turn their backs on a light source and face a wall. They perceive only the shadows (or projections) of the objects behind them. For them, these shadows are reality.

Remarkably, this allegory prefigures the metaphor of the cylinder, whose shadow is either circular or rectangular, depending on the orientation of the cylinder and the axis of the projection.

Figure 2 - Projected shadows of a cylinder

Let's Suppose an observer resides on the surface of the wall. In his vision of the world in two dimensions, this observer will interpret these two shadows as two distinct and separate objects in space. He cannot grasp the presence of the cylinder, which would require access to a third spatial dimension. If we change the orientation of the cylinder so that the rectangular shadow becomes circular, then the shadow that was circular automatically becomes rectangular. If the distances between the cylinder and the two shadows are equal, then this transformation of the two shapes is simultaneous, regardless of the distance between them. One can thus say that these two shadows are entangled, in the sense that the shape (or state) of one instantaneously determines the shape (or state) of the other, without any communication between them being required. Quantum entanglement seems incompatible with objective reality, but it becomes possible if what we observe of the Universe is only the projection of a deeper reality.

From a metaphysical point of view, a spatiotemporal projection from infinity to the finite is an obligatory passage to free oneself from the conceptual pitfalls related to objective physical reality, be it finite or infinite. In his first antinomy of pure reason, Kant opposes the thesis of a finite world (in time and space) to the antithesis of an infinite world. The proof of the thesis rests on the impossibility of the antithesis and vice versa. In this sense, objective realism can be seen as a metaphysical impasse. Projected reality is the solution to this ontological paradox. Systemic cosmology distinguishes ontology from physics, but these two components must be compatible and complementary.

The Universe in Itself and the Universe to Oneself

You are here invited to distinguish the Universe in itself (supposedly infinite and therefore immeasurable) from its finite projection, the Universe to oneself, the one that is a matter of physics and can therefore in principle be measured by an observer. According to Kant, we know the world only through the prism of our mental structure. We interpret phenomena with the help of our senses and understanding. The thing-in-itself, beyond its phenomenal reality, we cannot know it.

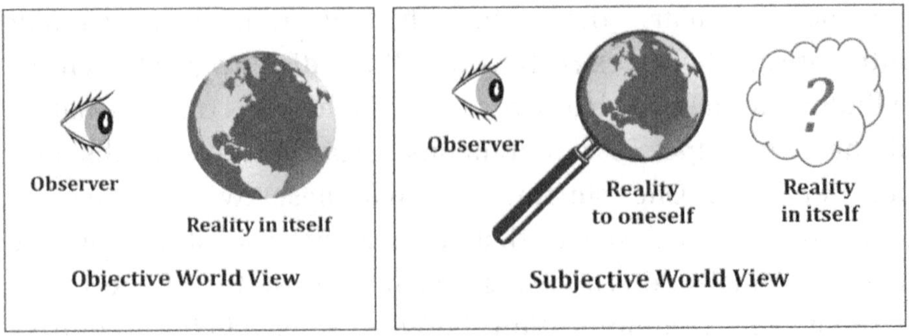

Figure 3 - Comparison between observational objectivity and subjectivity

This illustration is imperfect since the observer is himself part of the world he is observing. It aims to highlight this perceptual lens through which each observer apprehends the world differently. One can imagine that the perceptual lens is part of the observer's eye, and that the latter resides inside the cloud of reality in itself. The perspective effect, which is familiar to us, evokes this observational subjectivity. The same event or object is perceived differently depending on the point of view from which one observes it. Bernard d'Espagnat's rainbow is a good example. From a temporal point of view, Einstein showed that the notion of simultaneity was subjective. The order in which two events occur can be inverted for two observers who are distant or moving relative to each other.

This book tries to identify some properties of this perceptual lens, this projection screen which is interposed between the observer and reality in itself. What hinders the perception of this lens is partly a question of scale. For the relativistic effects induced by the maximum speed (c) to become perceptible, very high speeds must be involved. Similarly, it is only at a very small scale that the quantum effects induced by the minimum action (h) become apparent. The same principle would apply to the other minima and maxima of physical quantities. On a human scale, the cosmospheric effects induced by the maximum duration and maximum length would be perceptible. On a cosmological scale, they would become predominant. Our ability to objectify would therefore be limited both minimally and maximally. On the one hand, we have seen that the infinitely large and the infinitely small cannot exist objectively. On the other hand, the finite maximum of a physical quantity is by nature subjective.

The Flammarion wood engraving shows a pilgrim who crosses the firmament of an objectively closed world (figure 4). It illustrates the paradox of a universe objectively limited by a sphere of finite dimension.

Figure 4 - Paradoxical illustration of an objectively closed world

THE CONCEPT OF COSMOSPHERE

The Origin of the Concept of Cosmosphere

The origin of the concept of Cosmosphere is very old and dates at least to the time of the Eleatic philosophers (5th century BC).

Parmenides is considered the founder of ontology. In his poem *«On Nature»,* he postulates the existence of the One: the unique, eternal, unchanging, continuous and spherical being (the Cosmosphere).

Parmenides' cosmological and physical intuitions are surprisingly modern. At that time, it was generally believed that the earth was flat. Instead, Parmenides imagined a spherical earth divided into five climatic zones: two polar frozen zones; an equatorial torrid zone; and between these two extremes, two temperate and habitable zones. Although simplistic, this hypothesis of climatic zones associated with the intensity of the sunlight on earth seems to me innovative. For Parmenides, stars are celestial bodies, just like the moon and the sun. If the ancients had adopted the concept of the One of Parmenides, the evolution of science might have followed a different path.

For Plato, the Universe would rather have been forged by the demiurge, which constitutes a creationist cosmogony. In his cosmological treatise *«The Timaeus»*, we can read:

«This is why the god turned the world into a sphere, whose ends are everywhere at equal distance from the center, this circular shape being the most perfect of all and the most similar to itself... For nothing came out of it and nothing went in from anywhere since there was nothing outside of it.»

Unfortunately and wrongly, Plato assimilates the Cosmosphere to a starry sphere, and sets it in motion:

«He attributed to it a movement appropriate to its body, the one out of the seven movements that best suits intelligence and thought. Consequently, he made it rotate uniformly on itself at the same place and it is the circular movement that he imposed.»

This rotation of the celestial sphere was explaining the rotation of the stars that were fixed there. The earth was then considered the center of this starry sphere.

The Evolution of the Concept of Cosmosphere

In the Middle Ages, the Cosmosphere took on a more subtle form. The Book of the Twenty-Four Philosophers is a hermetic, anonymous work whose date of writing is uncertain. The second sentence of the book reads:

«God is an infinite sphere, whose center is everywhere and whose circumference is nowhere» (Hudry, 1989).

This concept of an infinite and intelligible sphere will be taken up by many authors including Alain de Lille, Meister Eckhart, Montaigne, Rabelais, Giordano Bruno and Blaise Pascal. Nicholas of Cusa also evokes it by avoiding the qualifier infinite. He rather assimilates the maximum sphere to the first mover of the world:

«Thus, the machine of the world will have, so to speak, its center everywhere and its circumference nowhere.» (Cues, Docte ignorance, 2008, p. 165)

Nicholas of Cusa has clearly freed himself from the sphere of fixed stars and circular orbits. Why then, maintain this concept of a maximum sphere, supposedly unique, uncreated, eternal, immutable and at first sight, not very compatible with the Scriptures?

It is because the *«maximum sphere»* is the One of Parmenides, the first principle from which everything derives, the first cause of the Universe and all that it contains:

« ... the learned ignorance sees very clearly that the absolute maximum is necessary, so that it is the absolute necessity. Now it has been shown that the Absolute Maximum can only be the One. Therefore, it is true that the maximum is the One.» (Cues, Docte ignorance, 2008, p. 52)

The Cosmosphere From an Ontological Point of View

In a way, cosmology cannot completely evacuate the concept of first cause. Nowadays, a big bang is invoked which suggests that time, space, energy and matter would have suddenly emerged from nothing. However, the ontology introduced by Parmenides teaches us that only being exists. Nothingness does not exist. Nothing can therefore emanate from that which does not exist. To get around this paradox, some models suggest a cyclical universe. For example, the Big Bang would only be the rebound of an ultra-dense state of matter/energy in a pre-existing space-time, and a big bang/big crunch cycle would repeat itself indefinitely. It is beneficial that different hypotheses are considered. However, it must be kept in mind that cyclical scenarios cannot explain the origin of the cosmos.

Since antiquity, philosophers have been telling us that the first cause cannot have a beginning, because then it should have a cause itself. Just as the first cause cannot be located in time, neither can it be located in space, because then a pre-existing space would be required for its very existence. The maximum sphere, imagined by Nicholas of Cusa, respects these precepts. It would constitute the limit of space and time, without however being part of it: it would thus be nowhere. Like an infinite sphere, its center can be moved without affecting it: it would therefore be everywhere.

Nicholas of Cusa also assimilates the maximum to God, who is the greatest, and for whom the qualifier of infinity is then reserved. Like his predecessor, Meister Eckhart, he will have to defend himself from serious accusations of pantheism. These accusations concern the principle of coincidence of opposites, and the concept of Cosmosphere of the Learned Ignorance:

«Others, who have tried to portray infinite Unity, have said that God is an infinite circle. And those who have considered the absolutely actual existence of God have said that God is like an infinite sphere. As for us, we will show that all these thinkers unanimously had a correct conception of the maximum, and that they all expressed a single doctrine.» (Cues, Docte ignorance, 2008, p. 67)

This ontology can be summarized as follows: The Universe being physically finite, it must be able to be contained. This cosmic and spherical container, Parmenides calls it *«The One»*, the hermetic Neoplatonism, *«the infinite sphere»* and the learned ignorance, *«the maximum sphere»*. The neologism of *«Cosmosphere»* is only the modern name for the evolution of this concept.

The Ontological Status of Time and Space

For the Cosmosphere to be invariant in time and space, it is necessary to redefine the notions of time and space in a cosmospheric context. In his *«Critique of Pure Reason»*, Immanuel Kant elaborates a transcendental theory of elements. According to Kant, time and space are the pure forms of sentient intuition, i.e. intuitive concepts that emanate from understanding, but which are in no way related to the senses:

«This research will result in the fact that there are two pure forms of sentient intuition, as a principle of a priori knowledge, namely space and time, which we will now examine.» (Kant, 1944, pp. 54-55)

Paradoxically, in the Kantian sense, it is the Cosmosphere that would be the a priori form of sentient intuition, and not time and space, as Kant believed. This is why Parmenides was able to imagine it, 25 centuries ago, in the absence of any conclusive observation. As these observations are now available, the Cosmosphere can now make a return into the cosmological arena. The cosmospheric principle postulates that it is the Cosmosphere that governs time, space, matter and motion. If we consider time and space as derived entities, then the apparent contradictions fade away.

The Cosmosphere From a Systemic Point of View

From a systemic point of view, there is nothing esoteric about the cosmospheric hypothesis. Every system operates and interacts in an environment that can be considered as a larger system. For example, the earth-moon system is part of the solar system, which is part of the Galaxy and so on. If we consider the Universe as a whole, then the question of its environment arises spontaneously. A global and autonomous system must have a self-containment mechanism. This mechanism is an integral part of the system and constitutes the essence of this book.

The Cosmosphere must be able to generate not only time, space, matter and movement, but also evolution, complexity, life and thought. Such a system remains difficult to describe in purely physical terms. Some philosophers have studied this cosmic evolution. In his book *«Le phénomène humain» (The Phenomenon of Man),* the scientist and theologian Pierre Teilhard de Chardin breaks new ground in this area (Chardin, 1955). Closer to us, the philosopher and theologian Jean Proulx, whose several works are listed in the bibliography, deals with this subject remarkably (Proulx, 2009).

The concept of Cosmosphere allows explaining several phenomena, today poorly understood. It has likely been discarded, partly because it has been falsely assimilated to the sphere of fixed stars, but also because it is poorly compatible with those of objective space and time, whether those of Descartes, Newton or even Einstein's space-time. Moreover, as an absolute maximum, the Cosmosphere would no longer be governed by the principle of non-contradiction. It would be the archetype of the coincidence of opposites. Whether we look in one direction or the opposite direction, it is the same Cosmosphere that surrounds us, no matter where or when.

From a systemic point of view, the Cosmosphere is the keystone, the missing centerpiece to which the other pieces of the cosmological puzzle can be attached to form a coherent whole. This simple, natural and aesthetic notion has fascinated the human mind for millennia. By giving up the concept of Cosmosphere, the baby has been thrown out with the bathwater, and it becomes very difficult to form a simple and coherent image of the Universe.

The Age of the Universe and Its Expansion Rate

Unlike the objects it contains, the Cosmosphere would not age. The maximum duration would be the age of the universe, about 13.578 Ga (gigayears). We have seen that the maximum of a quantity and the minimum of the inverse quantity are only two facets of the same entity. For example, if time is limited by a maximum duration, then a frequency must be minimally limited by the inverse of that duration. Indeed, an event subjected to a frequency lower than this minimum could never occur, since it would have to wait longer than the maximum duration. The same applies to a rate, whose dimension is the same as that of a frequency (T^{-1}). We conclude that the minimum rate should be the inverse of the age of the universe (the maximum duration).

Now, this is exactly what we observe. The minimum rate is none other than the expansion rate of the Universe, whose value is the Hubble constant, denoted H_0. In other words, we observe that the age of the Universe corresponds to what we now call Hubble time $t_H = 1/H_0$. The cosmospheric interpretation of this relationship is that just like the age of the universe, the Hubble constant is invariant. So I simply note it H, and the age of the Universe H^{-1} (or $1/H$).

According to conventional theories, this inverse relationship should not exist. The expansion rate of the Universe would vary over time according to the energy density associated with the different components of the Cosmos (matter, radiation, neutrinos, dark energy ...).

The inverse relationship between the maximum duration and the minimum rate is not the result of an unlikely coincidence, but rather the confirmation of previously formulated hypotheses. It is observational evidence of the existence of the Cosmosphere and its invariance.

The Radius of the Universe and the Maximum Speed

The maximum length would be the radius of the Universe that I note R_U. The Hubble radius being defined as $R_H = c/H$, it would also be the radius of the Cosmosphere. So I pose $R_U = R_H$ by definition. Physically, the Cosmosphere would correspond to what is called the Hubble sphere, with the difference that nothing physical would exist outside of it. The rest of this book shows that, despite appearances, the invariance of H and R_U is perfectly coherent. In a conceptual approach, it is important to distinguish between observation and its interpretation.

Regarding the maximum speed (c), we can see that its value is just enough to observe the first cosmic events such as the formation of primordial structures. Conventional theories must involve superluminal velocities to explain the isotropy of the primordial universe, or the existence of hypothetical galaxies floating beyond the cosmological horizon. According to the cosmospheric principle, it is normal to observe the relation $c = H R_U$. The use of superluminal velocities is neither required nor allowed.

THE COSMOSPHERIC PRINCIPLE

The cosmospheric principle explains the containment mechanism of the physical universe using the concept of Cosmosphere. It is a physical principle based on the uniqueness and invariance of an entity both transcendent and immanent to the world. Some assertions of Nicholas of Cusa suggest that he anticipated the cosmospheric principle, without, however, formulating it:

«Il is necessary that the maximum is in act the principle and the end of all finite things. Moreover, nothing could be if the Absolute Maximum were not. Indeed, since all things that are not the maximum are finite and dependent, it will be necessary that they come from another, because if they came from themselves, they would have been before they were. Now, as the rule shows, it is not possible to go to infinity in principles and causes. There will therefore be an Absolute Maximum without which nothing can be.» (Cues, Docte ignorance, 2008, p. 51)

«We now see clearly, since the Maximum is like the maximum sphere, how it is the only, the simplest and the most adequate measure of the whole universe and of everything that exists in the universe.» (Cues, Docte ignorance, 2008, p. 98)

The Formulation of the Cosmospheric Principle

The cosmospheric principle posits that it is the Cosmosphere that governs time, space and motion for any observer. The measurements associated with these quantities would therefore be subjective. The universe to oneself would take the appearance of a very large ball. This ball would be circumscribed by a sphere of dimension 2 which delimits time and space for any observer:

- **Anytime and anywhere, any massive body remains centered relative to the Cosmosphere, which is unique, invariant and therefore the same for all observers.**

Time and space measurements for each observer must therefore be adjusted to respect this principle. The invariance of the Cosmosphere is based on an ontological principle: something subject to change cannot exist by itself. The Cosmosphere cannot, therefore, change since it is itself the first cause, the Absolute Maximum of the Learned Ignorance. The uniqueness of the Cosmosphere is explicit since the Absolute Maximum is the One.

The cosmospheric principle differs from the Copernican (or cosmological) principle which postulates the isotropy and homogeneity of space. As mentioned above, if we do not have access to the Universe in itself, we can no longer formulate this kind of hypothesis. Rather, the principle adopted must apply to the universe to oneself, the one that can be observed and measured.

According to the cosmospheric principle, what is perfectly homogeneous and isotropic is the Cosmosphere itself. What is common to every observer is neither time, space nor its content, but simply the Cosmosphere. An objective of this book is to explain how and why this is not only possible, but also required for the understanding of several phenomena.

The Cosmospheric Paradigm

A paradigm is a way of looking at things, a coherent model of the world that rests on a defined basis. The cosmospheric principle implies a different paradigm from that of the Big Bang, which is widespread today. The comparison can be summarized as follows:

The Big Bang Paradigm (Objective Cosmology)	***The Cosmospheric Paradigm (Systemic Cosmology)***
The Universe is an object. It is ruled by the Copernican principle.	The Universe is a system. It is ruled by the Cosmospheric principle.
There is an objective reality that can be observed. This reality is independent of any observer.	Our reality is partly subjective. Thanks to observation and the concepts we forge, we can apprehend it.
The Universe in itself and the Universe to oneself are one and the same.	The Universe to oneself is a projection of the Universe in itself. This projection depends on the context of observation.

The Big Bang Paradigm (Objective Cosmology)	***The Cosmospheric Paradigm (Systemic Cosmology)***
The Universe is not spherical; it has neither center nor edge.	The universe to oneself is spherical and finite; it is centered on any observer.
The age of the Universe is about 13.8 Ga (gigayears). We can observe only a tiny part of it, and we do not know if it is spatially finite or not.	The age of the Universe is about 13.578 Ga (gigayears). Its radius is about 13.578 Gly (gigalight-years).
The primary cause of time, space, energy and matter is the Big Bang.	The primary cause of time, space, energy and matter is the Cosmosphere.
Space expands at the rate H_0, but time flows evenly.	Time and space dilate at a constant rate (H), which is the inverse of the age of the Universe. The age and the radius of the Universe (in light distance) are invariant.

The Illusion of the Big Bang

The Big Bang hypothesis results naturally from the adopted philosophical point of view: the objective realism from which the Copernican principle derives. It assumes a uniform flow of time as well as the homogeneity and isotropy of space. The cosmospheric interpretation of this phenomenon is that the Big Bang is only an optical illusion resulting from the compression of time and space in the past.

I believe that the adoption of cosmological hypotheses should be the result of a structured approach, and not of an arbitrary decision based on unproven beliefs. Presuming a priori the homogeneity of space and the uniform flow of time is, in my opinion, a form of cognitive bias. There is no reason to favor one point of view over the other. One should therefore rely on observation and reasoning before making up his mind. Discernment tells us that the paradigm that provides the best explanation of phenomena should be preferred.

COSMOSPHERIC INTERPRETATION OF TIME

Since antiquity, time has been the subject of many questions and different interpretations. There are countless books on the subject. Time plays a primordial role in any cosmological approach, and some questions inevitably arise such as:

- Has the Universe had a beginning and if so, what caused it? What was there before this beginning?
- Will the Universe come to an end and if so, why and how? What will happen after?

Modern Cosmology gives us a precious clue about the evolution of the cosmos. We now know that the Universe has a history. From a simple and monotonous past, it has gradually become structured, complex and organized. However, this does not answer fundamental questions about its origin and destiny. The cosmospheric point of view can be summarized as follows:

- Although its age is finite, the Universe has never had a beginning as such, because this age is invariant.
- Although it is continually evolving and changing, the Universe will never end.

Supposing that the Universe has had a beginning or that it will have an end is, in my opinion, a paradox, because the very notion of time prohibits the absence of a before and an after.

The Arrow of Time

The expression «Arrow of time» refers to the fact that time always seems to flow in the same direction: from the past to the future. Among the starting hypotheses, one must position oneself on the very existence of this arrow, its nature, its causes and its implications. It is generally accepted that past events are irreversible and that future events are not totally determined. For example, when approaching a junction, one can choose to oblique either to the left or to the right (in the immediate future). Once the action has been performed, it seems impossible to go back in time and change one's initial decision. From a physical point of view, the arrow of time is confirmed by the observation of the increase in entropy, a quantity associated with particle disorder. The expansion of the Universe also implies some form of temporal asymmetry. Also, the very concept of evolution implies that of an arrow of time, provided that one can assign a direction to this evolution, for example from simple to complex. Now, this is exactly what we observe.

Despite this evidence, many refute the existence of the arrow of time, on the pretext that the equations of physics allow an inversion of this arrow. The arrow of time would be only psychological. It is normal for opinions to diverge, but it is important to be aware of the implications of our own beliefs. If the arrow of time is just an illusion, then so are the decisions we believe we are making. If the past, present and future already coexist in a predetermined space-time, and that we are only witnessing events that happen throughout our lives, how could we influence these events? If the future is already determined, how could we intervene on this future?

Free Will

Free will is the ability to freely make certain decisions and to act accordingly to influence the course of events. From a metaphysical point of view, this seems obvious to me. If we have no control over our decisions and actions, why reward good deeds and punish the bad ones ? Ultimately, absolute determinism and the resulting denial of free will is nonsense. Indeed, this very belief would not result from the free choice of a thinking being, but simply from the interaction of a human brain with its environment. This pre-programmed belief would therefore have no argumentative value. To give up the right to choose our beliefs is to condemn ourselves to absurdity.

To achieve the desired result, we select and apply a sequence of appropriate and acquired behaviors, in interaction with our sensory perceptions. If the state changes were purely by chance, then a complex process such as driving a motor vehicle would be impossible. Our muscles would move in all directions and no coordination between our sensory perceptions and motor skills would be possible. We would not be able to choose a destination, let alone reach it. This ability to be able to choose and apply relevant changes of state would be what allows living beings to bypass the increasing entropy of a system. For a bird to fly and migrate to its destination, it must have a mechanism like the one described above. A robot that drives or flies can be made, but it too will have to simulate this mechanism.

In conclusion, the hypothesis put forward in this book is that, unlike the past, the future is not totally determined and that we can influence it. We, therefore, live in a universe where time is asymmetrical, or if you prefer, where the past is fundamentally different from the future.

Time Dilation

To explain this temporal asymmetry, the cosmospheric hypothesis posits that the flow of time is not constant, but rather that it is flowing more and more slowly. Previously, we have seen that time is not additive. The fact that time is adjunctive implies that time dilates at the minimum rate (H), so that the maximum duration (H^{-1}) is invariant. Consider the duration of one earth year, for example. The adjunction ensures that the years do not add up, but accumulate differently so that the result of this accumulation is slightly less than the sum of the elapsing years. If the current year is assigned the duration of one year, then this means that comparatively speaking, the years accumulated in the past are shorter. The older a year is, the shorter it is relative to the current year. Similarly, the coming year will be longer than the current year. This is the explanation for time dilation.

Measurement of Time

To measure time, a clock is used, and to know whether this clock runs too fast or too slow, we usually refer to a more accurate clock. But how can we know if the rate of the most accurate clocks is really constant? How to measure a possible variation in this rate?

It is not easy to answer these questions, and the very concept of time is difficult to grasp. Time is defined in relation to one or more clocks whose rate is assumed to be constant. The second is currently defined as the duration of 9 192 631 770 periods of the radiation matching the transition between two hyperfine levels of the ground state of the cesium 133 atom.

What matters here is that this duration is related to an atomic transition and therefore defined in relation to a property of matter. This is what I call a matter clock. What the rate of this matter clock could vary from?

Could the Light Clock Enlighten Us?

We can also imagine a light clock (or photonic clock), whose rate would be determined by the oscillation of a light wave. For example, if we observe the diffuse cosmological background and ignore our own motion, it seems isotropic. We then observe an important fossil radiation with a frequency of 160 GHz, which corresponds to a wavelength of 1.06 mm (microwaves). Compared to the matter clock, we observe that the rate of this light clock slows down very slightly ($H \approx 2.333\,8 \times 10^{-18}$ s^{-1}). This observation is asymmetrical, i.e. compared to the light clock, the matter clock accelerates very slightly.

Since we have two clocks that measure time differently, we may wonder why. Would one of these clocks measure time correctly the other not? Should we consider that the light clock is slowing down, or rather that it is the matter clock that is speeding up?

The Temporal Paradox

Fossil radiation was emitted a long, long time ago, and it is about a thousand times less energetic than if it were emitted today. According to the matter clock, 99.9 % of the energy of fossil radiation would have vanished! As early as 1936, Edwin Hubble evoked this problem of the non-conservation of photon energy in an expanding universe. It is therefore a known problem for which no satisfactory explanation has been provided.

Indeed, Noether's theorem demonstrates the equivalence between invariance by time translation and energy conservation. In other words, if the flow of time is uniform, then the energy produced in the past is conserved indefinitely.

This theorem allows formulating a deductive reasoning:

- *Premise:* A uniform flow of time implies that the energy produced in the past is conserved indefinitely.
- *Premise:* The matter clock indicates that the energy of photons emitted in the past is not conserved.
- *Conclusion:* The flow of time according to the matter clock is not uniform.

This temporal paradox is the Achilles' heel of cosmological modeling. Indeed, light constitutes almost all the information available on the Cosmos and its evolution. An erroneous interpretation of the cosmological spectral shift therefore results in the inconsistency of the model.

The Solution to the Temporal Paradox

We know that the energy of a photon is determined by its frequency f. If the frequency (f_S) measured by the emitting source differs from that measured by the observer (f_O), it is because the observer's clock does not beat at the same rate as that of the emitting source.

Fortunately, there is a Noether's time, the one for which the energy of any emitted photon is preserved indefinitely. This time is that of the light clock because it is the very frequency of photons that cadences this clock. The time of the light clock is therefore that in relation to which the laws of nature are invariant. For example, the speed of light would be constant according to the time of the light clock. Indeed, it is light that links time and space, through its speed, not matter.

If we perceive a dilation of the light time, it is because our matter clock accelerates in relation to the light clock. According to the light clock, the age of the Universe is invariant. This solution to the temporal paradox is corroborated by the observation of the expansion rate of the Universe, which corresponds well to the inverse of the age of the Universe.

COSMOSPHERIC INTERPRETATION OF SPACE

Space dilation

If the age of the Universe is invariant, then so is its radius in light distance. The rate of expansion (*H*) is therefore the same for these two adjunctive quantities. As with the unit of time, it is the very unit of length that dilates. Because the unit of time «year» dilates continuously, the unit of distance «light-year» does the same. This space-time dilation is the cosmospheric interpretation of the expansion of the Universe. It would allow an abundance of possibilities for the future. It would also be the explanation of the arrow of time: A unique and unchanging past, but a partially indeterminate future.

The Light Ruler and the Spatial Invariance of the Cosmosphere

Just as for the matter clock and the light clock, we can imagine a matter ruler and a light ruler. The matter ruler would be made of ordinary matter such as wood or metal. The light ruler would rather use the wavelength of the light rays of the light clock. As mentioned above, this wavelength would lengthen with respect to the matter ruler. If we interpret this phenomenon as an expansion of space, it is because we use the matter ruler to measure distances. From a light point of view, it would rather be our matter ruler and ourselves that shrink with respect to the invariant dimensions of the Cosmosphere.

COSMOSPHERIC INTERPRETATION OF MASS

We define mass as a measure of the quantity of matter, and matter as any substance that has mass. This tautology does not hide our difficulty to grasp these intuitive concepts.

Like time and space, mass would be a quantized and adjunctive quantity. If you buy two whole cheese wheels of one kilogram each, the combined mass of the two wheels will be slightly less than two kilograms. However, once you have eaten one kilogram of cheese, you will have another kilogram left over. If this seems strange to you, consider that you can add and subtract wheels of cheese, but not kilograms:

- 1 wheel + 1 wheel = 2 wheels, but
- 1 kg $\tilde{+}$ 1 kg < 2 kg.

The measurement of mass is intimately linked to the clock we use. In his 1924 thesis *«Recherches sur la théorie des quanta» (Research on the theory of Quanta),* Louis de Broglie broadened the interpretation of the Planck-Einstein relationship which was then limited to electromagnetic phenomena. It assumes the existence of a periodic phenomenon associated with mass:

«One can thus conceive that because of a great law of Nature, to each piece of energy of proper mass m, is linked a periodic phenomenon of frequency ν *such as we get:* $h\,\nu = m\,c^2$*. This hypothesis is the basis of our system.»* (Broglie, 1924)

This hypothesis was quickly adopted by the scientific community and confirmed by the electron diffraction in 1927.

The cosmospheric interpretation of this periodic phenomenon is the high frequency recentering of massive particles with respect to the Cosmosphere. This periodic and quantized recentering of particles is a direct result of the cosmospheric principle, which requires that massive bodies remain centered with respect to the Cosmosphere. This process would take place in a discontinuous manner due to the minimum action (h), all in accordance with the principle of least action. Each type of particle might correspond to a particular form of cosmospheric projection.

We can consider that a particle is actualized only during this recentering with respect to the Cosmosphere. Apart from these periodic events, the existence of the particle would only be potential, as described by the wave function of the quantum system of which it is a part, i.e. the entire Universe.

The concept of matter would thus be an emergent notion that results from multiple and incessant cosmospheric point projections, which exist side by side in time and space. The mass would correspond to the quantity of energy trapped in the volume of space where this process operates, i.e. the one that matter occupies.

This cosmospheric interpretation could contribute to a better understanding of certain quantum phenomena such as the problem of measurement.

The Increase in Mass

If the matter clock is accelerating, then the frequency f associated with a mass m is continuously increasing. Assuming h and c constant, we calculate $\dot{m} = H\,m$, since $h\,f = m\,c^2$. It is this increase in mass that would be responsible for gravity, not the mass itself. This variation being tiny, that would explain why the gravitational force is so weak. However, when the mass is large, as in the case of a planet, this effect would be clearly noticeable. For example, we know that the earth mass is $M_T \approx 5.972 \times 10^{24}$ kg. The increase in earth mass ($\dot{M}_T = H\,M_T$) would be about 14 000 t/s. For comparison, the mass of the steel structure of the Eiffel Tower is approximately 7 300 t.

The increase in mass is a hypothesis considered by many authors including (Jeans, 1928), (Dirac, 1937), (Hoyle, 1948), (Narlikar & Arp, 1993), (Fahr & Heyl, 2007) and (Wilhelm & Dwivedi, 2011).

The Dimension of the Gravitational Constant (G)

We can show otherwise that the mass of bodies is not constant. The gravitational constant (G) is a measure of the variation in space (of dimension L^3), induced by the presence of a mass (M). The very dimension of this constant tells us about the phenomenon. Let us suppose that the presence of a constant mass M is the cause of a variation in the surrounding space. This variation should also be constant, since at this scale, the same cause (mass M) induces the same effects (a variation in space). The dimension of the gravitational constant should then be $M^1\,L^3\,T^{-1}$, for example (m^3/s) per kg. According to the cosmospheric principle, to counteract this creation of additional space, the apple should then establish a constant speed in the direction of the earth.

However, this is not what we observe. Even assuming the gravitational field intensity to be constant, we do observe an acceleration of the apple, and consequently, the dimension of the gravitational constant is $[G] = M^1 L^3 T^{-2}$. When dropping an apple, there would be more space induced by the earth mass in the second second of the apple's fall than in the first second. There would be even more during the third second and so on, which seems paradoxical. If we drop an apple every second, after a few seconds, each apple will indicate a different measure of the variation in space for the same duration of fall.

In fact, what is constant is not the variation in space, but rather the variation in that variation. For this to be possible, a causal parameter must vary linearly over time, and this parameter can only be the mass M since no other parameter is involved.

To better understand the phenomenon, the unit of measurement of G can be interpreted as (m^3/s) per (kg s). According to this dimensional analysis, only a constant variation in the earth's mass can account for the constant acceleration of the apple.

Mass and Energy

Let us consider things from a cosmospheric point of view, that is we suppose that massive bodies contract with respect to the Cosmosphere. This contraction generates a maximum speed (c) of the bodies with respect to the Cosmosphere in all directions.

Since a body of mass M moves away from the Cosmosphere at speed c in one direction, it also moves away from it at the same speed in the opposite direction. The mass energy of this body ($E = M c^2$) thus seems to correspond to the kinetic energy it possesses with respect to the Cosmosphere.

THE COSMIC ACCELERATION

If time is dilating and the mass of bodies is increasing, it is because they are subject to an acceleration. All massive bodies, including matter clocks and ourselves would be accelerated relative to the Cosmosphere. This acceleration would be omnidirectional, and its value would correspond to the product of the minimum frequency (H) by the speed of light (c), i.e. $H\,c \cong 6.996\,5 \times 10^{-10}$ m/s^2. For comparison, the Earth's attraction is 1.4×10^{10} times more intense.

The implications of this cosmic acceleration ($H\,c$) are huge. The Big Bang paradigm suggests that time, space, matter and energy would have suddenly emerged from nothingness. The cosmospheric principle rather suggests that these entities are continuously being generated by the Cosmosphere, and that the Universe is in perpetual evolution. This cosmic evolution affects not only our reading of the past, but also that of our future, in the very long term.

If there were no cosmic acceleration, we would not observe this difference between the light clock and the matter clock. The time arrow would not exist and time would not be able to elapse. Everything would be frozen; no change, no movement and no evolution would be possible.

The Cosmospheric Compressive Force

One consequence of cosmic acceleration is that massive bodies would be caught in a vice by the Cosmosphere. They would be subjected to an omnidirectional compressive force. To explain the increase in the mass M of a body, it is assumed that a force $M H c$ applied in any direction on this body would be compensated by an equivalent force in the opposite direction.

The Pioneer Anomaly

These concepts of light clock, acceleration of atomic clocks and an increase in the mass of all massive bodies are seriously considered by recognized physicists. In 2011, K. Wilhelm (Max-Planck Institute for Solar System Research) and B. N. Dwivedi (Hindu University of Benares) published a paper that tries to explain an anomaly of the Pioneer 10 and 11 space probes. These seem to undergo a slight unexplained deceleration whose value is close to the cosmic acceleration ($H c$) described above. The hypothesis considered is an acceleration of the rate of the terrestrial atomic clocks, which results in an increase in the mass of all massive bodies:

«Based on a gravitational impact model - requiring a secular mass increase of all massive bodies - a solution is proposed that implies a clock acceleration with a value close to that of the Hubble constant.» (Wilhelm & Dwivedi, An Explanation of the Pioneer anomaly involving accelerated atomic clocks, 2011.)

In accordance with the cosmospheric model, the value of the acceleration of atomic clocks would be close to that of the Hubble constant. Although other more conventional explanations are considered, the probability that they correspond to this value is tiny. In addition, similar anomalies have also been observed on the Galileo and Ulysses probes. Thanks to the increased accuracy of the new atomic clocks, cosmic acceleration could be measured over much shorter durations than those associated with the voyages of the Pioneer probes. Unfortunately, NASA and ESA seem to have abandoned plans to elucidate this anomaly.

The Non-Conservation of Energy and Momentum

In physics, a conservation law states that a measurable property of a physical system remains constant during the evolution of this system. On short times or distances, we can indeed consider that energy and momentum are conserved. However, we saw earlier that at the cosmological scale, conventional models imply that the energy of the emitted photons is not conserved.

We also saw that Noether's theorem shows that energy is conserved over time. However, this demonstration presumes that the flow of time (according to the matter clock) is uniform, which the cosmospheric hypothesis refutes.

One way or another, we conclude that energy is not conserved at the cosmological scale. Conventional models assume a loss of energy of the emitted photons while the cosmospheric model implies an increase in the energy of massive particles.

The same is true for momentum. This quantity would only be conserved if the translation in space were indeed invariant, i.e. if space were homogeneous, which is also refuted by the cosmospheric hypothesis.

Let us take the example of the Pioneer anomaly described above. This phenomenon can be interpreted spatially. From its point of view, the Pioneer 10 probe maintains a constant speed and no deceleration is felt. However, because it is moving away from us, the meters it travels seem shorter and shorter to us. This is why we perceive a deceleration of this probe. The momentum of this probe is therefore not conserved from the point of view of a terrestrial observer.

Now, what would happen if this probe turned around and came back to us at the same speed? Logically, the meters it travels would seem longer and longer to us and the speed of the probe would seem to increase.

Ultra-High Energy Cosmic Rays

Cosmic rays are particles of ionized matter (e.g. protons, helium nuclei, electrons...), many of which come from the sun. However, some of these particles are of extragalactic origin and reach an energy level higher than 10^{20} eV. Such energy is macroscopic and comparable to that of a well-hit tennis ball.

It is assumed that extreme magnetic fields are responsible for this phenomenon, but for the moment, the origin of these fields and the physical processes at work are unknown. An alternative or complementary explanation would be cosmic acceleration ($H\,c$). Under the effect of this acceleration, a particle of matter coming from the confines of the Universe could reach a relativistic speed, and possess a colossal kinetic energy.

THE EVOLUTION OF THE UNIVERSE: A QUESTION OF PRINCIPLE

To conceive a coherent cosmological system, it is essential to question the principles and assumptions on which this system relies. Systemic assumes the existence of guiding principles that govern the evolution of the cosmos. Among these, two extremum principles can be identified:

- A minimum principle (the principle of least action) that minimizes the occurrence of a minimum and invariant entity: The quantum of action.
- A maximum principle (the cosmospheric principle) that guarantees the uniqueness of a maximum and invariant entity: The Cosmosphere.

We previously saw that extremum values act like beacons. They allow containing physical quantities within finite measurement ranges, thus avoiding singularities. Guiding principles, on the other hand, allow streamlining and guiding the coherent evolution of the system. Without them, the Universe would be doomed to chaos, with no possibility of structuring, increasing complexity, or self-organization.

We know that a very long time ago, the Universe was in a state of thermal equilibrium, essentially made up of ionized hydrogen. Now, this physical state is characterized by maximum entropy, i.e. it is totally disorganized. How then can we reconcile the complex organization of living beings that we observe today with the global increase in entropy that was once already maximized ?

To bypass the increase in entropy, a system must have a source of energy that can do useful work. In the case of a plant, this energy source is solar radiation. For an animal it is food, and for a vehicle, it can be a fuel or a battery. In any case, this source of energy must be renewed.

The Universe being an evolutionary system, it must itself have an inexhaustible source of energy to be able to evolve. This energy comes from the Cosmosphere, which constantly communicates it to all particles of matter. This is why these particles continuously gain mass. Thanks to this renewed energy, matter can organize itself. Particles can arrange themselves to form increasingly complex structures: atoms, molecules, cells and multicellular organisms.

Without beacons, guiding principles, energy sources nor arrow of time, chance can only generate the most probable state: chaos. I conclude that objective cosmology can hardly account for the Universe we inhabit, its complexity and its evolution.

BEYOND OPPOSITES

Before approaching cosmological modeling as such, I think it is useful to clarify a point concerning *«the ambition of overcoming opposites»* that Werner Heisenberg evoked in the last century. The *«coincidence of opposites»* tells us that this overcoming of opposites applies to the maximum as well as to the minimum. It operates a certain form of conciliation between different points of view that seem at first sight irreconcilable. Can we really get around Aristotle's principle of non-contradiction, and if so, how?

Part of the answer to this question lies in the design process itself as described in the introduction. Things can be seen in diverse ways and these different points of view can lead to divergent conclusions. For example:

- The Universe in itself would be infinite while its projection, the Universe to oneself would be finite.
- According to the light clock, the age of the Universe would be invariant, while it seems to age according to our matter clock.
- According to the light ruler, the Cosmosphere's radius would be invariant, even if we observe that it expands at speed c according to our matter ruler.

In the text, I use different points of view depending on the context. For example, cosmic expansion implies the use of matter instrumentation, while cosmospheric invariance refers to the use of light instrumentation. There is therefore no real contradiction in the logical sense of the term.

THE COSMOSPHERIC MODEL

The next step is to identify and assemble the pieces of the cosmological puzzle. In the introduction, I mentioned that modeling essentially consists in identifying concepts, and making them interact to represent the coherent, efficient, simple and aesthetic functioning of the system.

The MELT-hc Network

To concretize this cosmic modeling, I assimilate the concepts to certain physical quantities (*M*: Mass, *E*: Energy, *L*: Length and *T*: Time), and the interactions to known formulas that link these quantities together:

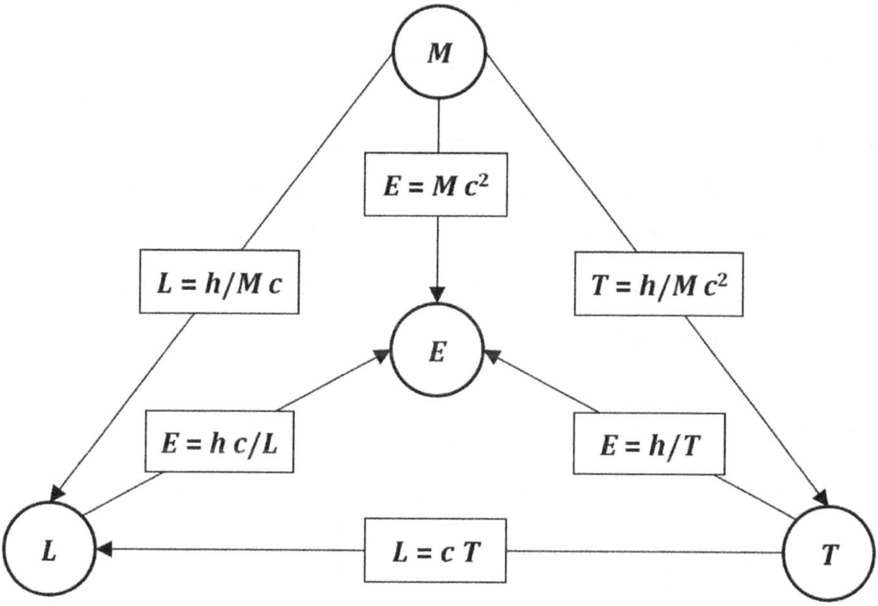

Figure 5 – The *MELT-hc* network

This kind of illustration is typically used by system designers, and a network-type structure is quickly detected. The physical quantities (*M*, *E*, *L* and *T*) are represented by circles, and the relationships by arrows, each with a boxed formula. Each arrow indicates the direction of the relationship expressed by the formula linking two quantities. This direction is arbitrary and can be reversed by changing the formula. For example, the relation $E = M c^2$, Einstein's famous formula can be written $M = E/c^2$.

The *MELT-hc* network shows known equations that express the wave nature of energy and mass. The value of *L* then corresponds to a wavelength, and that of *T*, to the period of this wave. In this model, the relation $L = c\ T$ links time to space in the same way that $E = M c^2$ links mass to energy. It is a direct consequence of the cosmospheric principle. In fact, all the relationships of the model have this same universal status.

The relation $E = h/T$ is a reformulation of the Planck-Einstein relation ($E = h f$), expressed as a function of time (*T*). The relation $L = h/M c$ is the Compton wavelength associated with a mass *M*, and so on. These relatively simple relationships can be deduced from each other. They involve two extremum values:

- the minimum action (*h*), and
- the maximum speed (*c*).

This explains the name of the network: *MELT-hc*.

The Cosmospheric Network

The *MELT-hc* network only illustrates a part of the cosmological puzzle. Of course, we can add to it other physical quantities such as Force (*F*) or Action (*A*). According to the cosmospheric model, all extremum values can be considered as universal constants. For simplicity and conformity, the usual constants (like *c*, *h* and *G*) are used. Since we have new hypotheses on the physical quantities, we can imagine the global picture of the puzzle:

- **A network of physical quantities linked together by algebraic equations involving the extremum values of other quantities.**

To be able to formulate these equations and calculate the extremum values, additional physical constants are required. If we limit ourselves to the above, there is not enough information to put all the pieces of the puzzle together.

In a first step, the gravitational constant (*G*) can be used: $L = 2G\,M/c^2$. This length *L* is the so-called Schwarzschild radius, i.e. the radius of a sphere below which nothing can escape the gravitational effect of a mass *M*, not even light. It is thus, in a sense, the minimum length that can be associated with a mass *M*.

An additional constant is required, and this is where the Hubble constant (*H*) comes into play. The observed value of this minimum rate (or minimum frequency) is about 72 km/s per megaparsec. This model uses a calculated value of H which is discussed in a subsequent section. This rate would be 72.013 km/s per megaparsec, which corresponds to a frequency of $2.3338 \times 10^{-18}\ s^{-1}$.

And voila, we now have all the elements required to assemble the cosmological puzzle.

Notation of Physical Quantities and Extremum Values

In addition to the physical quantities shown in Figure 5 (*M, E, L* and *T*), the following should be added:

- *f*: for frequency, i.e. the inverse of a duration of dimension T
- *F*: for Force
- *v*: for speed (or velocity)
- *a*: for acceleration
- *A*: for Action
- *Θ*: for Temperature

The subscript $_{\text{min}}$ refers to the minimum value of a quantity.

The subscript $_{\text{max}}$ refers to the maximum value of a quantity.

When an extremum value applies to the whole Universe, the subscript $_{\text{U}}$ is sometimes used.

Assembly of the Cosmological Puzzle

The assembly of the cosmological puzzle is an algebraic procedure that consists in calculating the extremum values from those already known and from the previously identified relations.

We already know that:

$v_{max} = c = 299\,792\,458$ m s^{-1}

$A_{min} = h = 6.626\,070\,15 \times 10^{-34}$ kg m^2 s^{-2}

$f_{min} = H \approx 2.333\,8 \times 10^{-18}$ s^{-1}; it is also the expansion rate of the Universe 72.013 km/s per megaparsec.

$T_{max} = H^{-1} \approx 4.284\,9 \times 10^{17}$ s , or 13.578 Ga (gigayears), the age of the Universe.

$L_{max} = c/H \approx 1.284\,6 \times 10^{26}$ m, or 13.578 Gly (gigalight-years), the radius of the universe, also noted R_U.

$a_{min} = H\,c \approx 6.996\,5 \times 10^{-10}$ m s^{-2}, the cosmic acceleration described above.

As $E = h\,f$, $E_{min} = h\,f_{min} = h\,H \approx 1.546\,4 \times 10^{-51}$ J

As $M = E/c^2$, $M_{min} = E_{min}/c^2 = h\,H/c^2 \approx 1.720\,6 \times 10^{-68}$ kg

As $L = 2G\,M/c^2$, $L_{min} = 2G\,M_{min}/c^2 = 2G\,h\,H/c^4 \approx 2.555\,5 \times 10^{-95}$ m

As $T = L/c$, $T_{min} = L_{min}/c = 2G\,h\,H/c^5 \approx 8.524\,1 \times 10^{-104}$ s

As $v = L/T$, $v_{min} = L_{min}/T_{max} = (2G\,h\,H/c^4)\;/\;H^{-1} = 2G\,h\,H^2/c^4$
$\approx 5.963\,9 \times 10^{-113}$ m s^{-1}

As $F = M\,a$, $F_{min} = M_{min}\,a_{min} = (h\,H/c^2)\,(H\,c) = h\,H^2/c$
$\approx 1.203\,8 \times 10^{-77}$ N

To calculate the maximum values*, we use the inverse relation:

$f_{max} = 1/T_{min} = c^5/2G\,h\,H \approx 1.173\,1 \times 10^{103}\ s^{-1}$

As $E = h\,f$, $E_{max} = h\,f_{max} = c^5/2G\,H \approx 7.773\,4 \times 10^{69}$ J, it is the global energy of the Universe.

As $M = E/c^2$, $M_{max} = E_{max}/c^2 = c^3/2G\,H \approx 8.649\,0 \times 10^{52}$ kg, the mass of the Universe also noted M_U

As $a = f\,v$, $a_{max} = f_{max}\,c = c^6/2G\,h\,H \approx 3.517\,0 \times 10^{111}\ m\ s^{-2}$

As $F = M\,a$, $F_{max} = M_{max}\,a_{max} = (c^3/2G\,H)\,(c^6/2G\,h\,H) = c^9/4G^2\,h\,H^2$ $\approx 3.041\,9 \times 10^{164}$ N

As $A = E\,T$, $A_{max} = E_{max}\,T_{max} = (c^5/2G\,H)\,(H^{-1}) = c^5/2G\,H^2$ $\approx 3.330\,8 \times 10^{87}$ kg m^2 s-2

Concerning the temperature (noted Θ), an order of magnitude of the extremum values Θ_{min} and Θ_{max} can be calculated using the Boltzmann constant ($k_B = 1.380\,649 \times 10^{-23}\ J\ K^{-1}$).

$\Theta_{min} \approx (E_{min}/k_B = h\,H/k_B) \approx 1.12 \times 10^{-28}$ K

$\Theta_{max} \approx (E_{max}/k_B = c^5/2G\,H\,k_B) \approx 5.63 \times 10^{92}$ K

These temperatures are expressed in terms of entropy and heat. They do not consider the number of degrees of freedom of the particles.

* Extremum values are listed in appendices B and C.

The Choice of Fundamental Constants

In the above equations, I used the conventional constants (c, h, G and k_B) to which I have added the Hubble constant (H). However, this choice is arbitrary, and the cosmospheric algebra allows the use of other values.

For example, one could consider that the gravitational constant G is derived from extremum values, considered more fundamental. Indeed, if $M_U = c^3/2G\,H$, it means that $\mathbf{G = c^3/2H\,M_U}$. By replacing the expression ($2G$) in the equations of the model by ($c^3/H\,M_U$), we obtain:

$L_{min} = h/\,M_U\,c$	instead of	$L_{min} = 2G\,h\,H/c^4$
$v_{min} = H\,h/M_U\,c$	instead of	$v_{min} = 2G\,h\,H^2/c^4$
$T_{min} = h/M_U\,c^2$	instead of	$T_{min} = 2G\,h\,H/c^5$
$E_{max} = M_U\,c^2$	instead of	$E_{max} = c^5/2G\,H$
$A_{max} = M_U\,c^2/H$	instead of	$A_{max} = c^5/2G\,H^2$
$f_{max} = M_U\,c^2/h$	instead of	$f_{max} = c^5/2G\,h\,H$
$a_{max} = M_U\,c^3/h$	instead of	$a_{max} = c^6/2G\,h\,H$
$F_{max} = M_U^2\,c^3/h$	instead of	$F_{max} = c^9/4G^2\,h\,H^2$

The use of M_U instead of G seems to me simpler and easier to interpret.

These equations aim to illustrate the possibilities of the cosmospheric model. However, for conformity, I generally use the conventional constants.

Cosmospheric Algebra

The relationships of the cosmospheric network can be expressed in many ways, using different extremum values. For example, to calculate the mass of the universe (M_U), one can use:

$M_{max} = E_{max}/c^2 = L_{max}\, c^2/2G = T_{max}\, c^3/2G = hc/2G\, M_{min}$
$= hc^3/2G\, E_{min} = h/c\, L_{min} = h/T_{min}\, c^2 = c^3/2G\, H$ ...

The pieces of the cosmospheric puzzle all fit together. If you are comfortable with algebra, you can formulate additional equations yourself using those already provided. If like me, you like intrigue, I suggest you look at **Appendix G - The *MELT-G* Enigma.**

An important relationship of the model arises from the extreme values of mass. Indeed, we can see that the product $M_{min}\, M_{max} = (h\, H/c^2)\,(c^3/2G\, H) = h\, c/2G$, from which $\boldsymbol{h\, c = 2G\, M_{min}\, M_{max}}$. This relation will be used later.

Further Calculation of Specific Values

If we assume that the minimum (s_{min}) and maximum (s_{max}) surfaces correspond to those of disks centered on the observer, then:

$s_{min} = \pi L_{min}^2 = \pi (2G h H/c^4)^2 = 4\pi G^2 h^2 H^2/c^8 \approx 2.051\,6 \times 10^{-189}$ m^2

$s_{max} = \pi L_{max}^2 = \pi c^2/H^2 \approx 5.184\,1 \times 10^{52}$ m^2

If we assume that the minimum (V_{min}) and maximum (V_{max}) volumes correspond to those of spheres centered on the observer, then:

$V_{min} = (4/3)\pi L_{min}^3 = (4/3)\pi (2G h H/c^4)^3 = 32\pi G^3 h^3 H^3/3c^{12}$
$\approx 6.990\,2 \times 10^{-284}$ m^3

$\boldsymbol{V_{max}} = (4/3)\pi L_{max}^3 = (4/3)\pi (c/H)^3 = \boldsymbol{4\pi c^3/3H^3}$
$\approx 8.879\,2 \times 10^{78}$ m^3, the volume of the Universe also noted $\boldsymbol{V_U}$.

As we previously calculated that $\boldsymbol{M_U = c^3/2G H}$, the density of the universe would be:

$M_U/V_U = (c^3/2G H) / (4\pi c^3/3H^3) = \boldsymbol{3H^2/8\pi G}$
$\approx 9.740\,8 \times 10^{-27}$ kg m^{-3}, and its energy density:

$\boldsymbol{\rho_U = 3c^2 H^2/8\pi G = \rho_C} \approx 8.754\,6 \times 10^{-10}$ J m^{-3}

We then find the Friedman equation that describes a universe whose density (ρ_U) is equal to the critical density (ρ_C). The cosmospheric model thus implies a universe of zero curvature according to general relativity, i.e. it would possess a Euclidean geometry on a large scale. To date, this result is corroborated by observation, and confirms the consistency of the cosmospheric model.

If we use M_U instead of G, the Friedman equation becomes:

$\boldsymbol{\rho_U = 3H^3 M_U/4\pi c}$, which logically corresponds to $\boldsymbol{\rho_U = M_U c^2/V_U}$.

The Entropy of the Universe (S_U)

Entropy can be viewed as a measure of the information required to describe the state of a system. By positing $k_B = 1$, entropy can be expressed in nat, which is the natural unit for quantifying this quantity. The global entropy of the Universe can then be associated with a ratio that I note (Σ):

$$\boldsymbol{S_{U(nat)} = \Sigma = E_{max}/E_{min}} = (c^5/2G\,H)/(h\,H) = \boldsymbol{c^5/2G\,h\,H^2}$$
$$\boldsymbol{\approx 5.026\,8 \times 10^{120}}$$

One could just as well calculate this mysterious ratio using another quantity such as Action or Frequency. Indeed, we can see that the max/min ratio of a physical quantity is a natural exponent of this dimensionless constant (Σ). For example, the max/min ratio of the quantities Length, Time and Mass is of the first degree (Σ). The max/min ratio of the quantity Surface is of the second degree (Σ^2) and the max/min ratio of the quantity Volume is of the third degree (Σ^3).

There are other ways to express entropy. The SI system recommends expressing entropy in joules/kelvin. The global entropy of the Universe is then: $S_{U(SI)} = \Sigma \times k_B = 6.940\,3 \times 10^{97}$ J/K.

The bit (binary digit) is also used as the unit of measure of entropy, and $S_{U(bit)} = \Sigma / \ln(2) = 7.252\,2 \times 10^{120}$ bits. This value then resembles the Lloyd limit which corresponds to the number of bits of information processed by the universe since its origin (Lloyd, 2002).

The Wesson's Mass and the Cosmological Constant (Λ)

In 2003, astrophysicist Paul S. Wesson, then a professor at the University of Waterloo, published an article entitled *«Is Mass quantized?»* It assumes the existence of a quantum of mass, that is, the smallest amount of mass that can physically exist.

This quantum of mass that Wesson notes m_P (in reference to Planck) is calculated using the constants c, h and Λ. The cosmological constant (Λ) is used by the cosmological models of type ΛCDM (Lambda Cold Dark Matter). It does not intervene in the calculation of the extremum values of the cosmospheric model, but strangely, Wesson's quantum mass corresponds to the minimum mass of the cosmospheric model:

$m_P \approx M_{min} \approx 2 \times 10^{68}$ kg.

According to Wesson, $m = (h/c)\,(\Lambda/3)^{1/2}$, compared to $M_{min} = H\,h/c^2$ according to the cosmospheric model. Multiplying these equations by c/h gives $(\Lambda/3)^{1/2} = H/c$, hence:

$\boldsymbol{\Lambda = 3H^2/c^2 \approx 1.818 \times 10^{-52}\ \mathrm{m}^{-2}}$, according to Wesson's Λ hypothesis.

Moreover, using the constants c, G and Λ, Wesson calculates the global mass of the universe which he notes m_E (in reference to Einstein): $m_E = (c^2/G)\,(3/\Lambda)^{1/2}$. If we disregard the factor 2 in the cosmospheric formulation ($M_U = c^3/2G\,H$), we find the relation $\Lambda = 3H^2/c^2$. Moreover, these two maximum masses are of the same order of magnitude: $m_E \approx M_U \approx 1 \times 10^{53}$ kg.

Professor Wesson concludes:

«In summary, astrophysical data indicate that we should add the cosmological constant Λ to the suite of fundamental physical parameters, which implies a mass related to Plank's constant of

approximately 2 × 10^{-65} g and a mass related to the gravitational constant of approximately 1×10^{56} g.» (Wesson, 2003)

The calculations of Professor Wesson, unfortunately deceased in 2015, corroborate the cosmospheric model and the hypotheses previously formulated on the extremum values of physical quantities. By adding a cosmological parameter (either H or Λ) to the so-called fundamental constants (c, h and G), dimensional analysis reveals the existence of extremum values for mass, and these values can be calculated.

The concordance of the two forms of calculation confirms that the rate of expansion of the Universe (H) is not just any value that varies over time, but rather a universal constant that allows calculating the extremum values of mass, and therefore those of energy, frequency, etc.

The Hubble Constant and Calculation of Mass

Using the Hubble constant to calculate extremum values of mass is not a new idea. In 1958, Fred Hoyle presented his Steady State Model at the Solvay Congress of Physics in Brussels (Hoyle, The Steady State Theory, 1958). The cosmospheric formulation of the mass contained in the Hubble sphere is then formulated: $\boldsymbol{M_U = c^3/2G\,H}$.

Similar equations were formulated by (Narlikar, 1993), (Carvalho, 1995) and (Haranas & Harney, 2009). In 2013, Dimitar Valev did the same and calculated a tiny mass of the same order of magnitude as the minimum mass: $m_1 \sim \hbar\, H/c^2$ (Valev, 2013).

Beyond the precision of the calculations, the cosmospheric model assimilates these extremum values to fundamental constants, and allows them to be embedded in a unified system.

Planck Units

Planck units are very small or very large values, which are calculated using so-called fundamental physical constants such as c, G, $\hbar$ ($h/2\pi$) and k_B (the Boltzmann constant). Applying the rules of dimensional analysis, we get the following values:

- Planck length: $L_P = \sqrt{G\hbar/c^3} \approx 1.616 \times 10^{-35}$ m
- Planck time: $T_P = \sqrt{G\hbar/c^5} \approx 5.391 \times 10^{-44}$ s
- Planck mass: $M_P = \sqrt{\hbar c/G} \approx 2.176 \times 10^{-8}$ kg
- Planck energy: $E_P = \sqrt{\hbar c^5/G} \approx 1.956 \times 10^9$ J
- Planck temperature: $\Theta_P = E_P/k_B \approx 1.417 \times 10^{32}$ K

According to the cosmospheric model, these values do not correspond to the extremum values of the listed quantities.

For example, the Planck mass ($M_P \approx 2.176 \times 10^{-8}$ kg) is about the same as that of a tiny insect. This may be the minimum mass for an insect, but certainly not for an elementary particle like the electron, which is about 9.109×10^{-31} kg.

The Planck energy ($E_P \approx 1.956 \times 10^9$ J) is approximately 278 kWh, which corresponds to the energy released by 31 L of gasoline. This value is therefore not a physical limit.

To calculate extremum values other than c and h, an additional constant is required, such as H, for example.

Intermediate Values of the Cosmospheric Model

Intermediate values* of the cosmospheric model (subscript **int**) are of the same order of magnitude as Planck units. They correspond to the square root of the product of the extremum values.

For example, $M_{int} = \sqrt{M_{min} M_{max}} \approx 1.77\ M_P$ is the mass for which the Compton wavelength is equal to the Schwarzschild radius:

$L_{int} = \sqrt{L_{min} L_{max}} \approx 3.54\ L_P$.

These values also have the particularity of being equivalent to the inverse intermediate value. For example, $T_{int} = 1/f_{int}$ and $f_{int} = 1/T_{int}$.

The difference between the intermediate values of the cosmospheric model and the Planck units is due, on the one hand, to the use of a factor 2 associated with the constant G, and, on the other hand, to the use of h instead of $\hbar$. On this last point, the cosmospheric system does not allow the use of an action value that would be lower than the minimum action (h). Of course, one can use the expression $h/2\pi$ in an equation, but this value is then an action/radian. Since the radian is an angle (a dimensionless quantity), the use of $\hbar$ may give the impression of an action quantity smaller than h.

* Intermediate values are listed in appendix D.

Is the Universe a Black Hole?

Mathematically, the cosmospheric model assumes that the universe is a gigantic cosmic black hole of mass M_U, and whose horizon is the Cosmosphere ($R_U = 2G\ M_U/c^2$). As we have seen previously, the Universe is a closed system, closed on itself. It is unthinkable that anything could cross the Cosmosphere, even a single photon. The confinement of the Universe by the Cosmosphere is therefore compatible with the physical properties of the cosmic black hole.

However, we must distinguish the Universe from the black holes it contains. Stellar or galactic black holes were forged by matter. They have a birth, a growth, and if we refer to the present theories, they will eventually evaporate. They rotate on themselves, can be electrically charged, and grow by swallowing the surrounding matter.

As for the Universe as a whole, there is none of that. As Plato said twenty-four centuries ago, *«Nothing came out of it and nothing came in from anywhere, since there was nothing outside of it.»* The Cosmosphere is indeed the horizon of the Universe, but unlike the horizon of ordinary black holes, it is invariant.

You may wonder what the inside of a black hole looks like. What happens at the center of a black hole, where theoretically the curvature of space-time becomes infinite?

The cosmospheric hypothesis is amazing. It states that we already occupy the center of the cosmic black hole, and that whatever we do, we are destined to remain there.

The Factor 2 Preceding the Gravitational Constant G

The relation $L = 2G\,M/c^2$ (the Schwarzschild radius of a mass M) implies that a factor 2 preceding the gravitational constant G is present in the equations of the model. For example:

$\boldsymbol{F_{min} = 2G\,M_{min}\,M_{max}/L_{max}^2}$ and $\boldsymbol{F_{max} = 2G\,M_{min}\,M_{max}/L_{min}}$.

This factor 2 can be derived from the kinetic energy (E_k) of a body of mass m moving at speed v ($E_k = ½m\,v^2$). Indeed, the gravitational force is said to be conservative, that is to say that the sum of the kinetic energy (E_k) and the potential energy (E_p), is conserved.

To escape the gravitational pull of a body of mass M at a distance L, the kinetic energy of the moving body must be equal to (or greater than) the potential energy (E_p) exerted by the body of mass M at that distance ($E_p = G\,M\,m/L$).

By positing $E_k = E_p$, we obtain $L = 2G\,M/v^2$, where v is the escape velocity of the moving body at distance L. If we maximize the velocity ($v = c$), we then minimize L as a function of M, which corresponds to the Schwarzschild radius.

In fact, this factor 2 is also taken into account in the relativistic equations of Einstein and Friedman.

THE ACCURACY OF THE EXTREMUM VALUES OF THE MODEL

The measured value of the rate of expansion of the Universe (H) is so far imprecise. Observed values range from 70 to 74 km/s per megaparsec, depending on the techniques used. The calculated value of the cosmospheric model should therefore suffer from this imprecision. To fill this gap, another known but more accurate value must be used. This increased accuracy then propagates to the value of H and the other calculated values of the model. A potential candidate would be the elementary charge (noted e). Since May 2019, the value of the elementary charge is set by definition to $1.602\,176\,634 \times 10^{-19}$ C.

The Coulomb's Law

Coulomb's law governs the electrostatic force that results from the interaction between two charges. It has a form similar to the law of universal gravitation: $F = k_e q_1 q_2 / r^2$, where k_e is the Coulomb constant ($k_e \approx 8.987\,551\,792 \times 10^9$ N m^2 C^{-2}), q_1 and q_2 are two interacting charges expressed in coulombs, and r is the distance in meters between these two charges considered as point charges.

Like the gravitational force, the electrostatic force is conservative, and a factor of 2 sometimes precedes the Coulomb constant (k_e) for the same reasons described above.

The Fine Structure Constant (α)

The fine structure constant (noted α) is a dimensionless quantity also called the coupling constant for the electromagnetic interaction. It corresponds to the ratio $k_e\, e^2/\hbar\, c$ and its value is about 1/137. Considering the observations previously made on the use of ħ and k_e, the ratio we are looking for would rather be $2k_e\, e^2/h\, c$, or α/π.

As we saw previously that $hc = 2G\, M_{min}\, M_{max}$, we deduce that

$\boldsymbol{\alpha/\pi = k_e\, e^2/G\, M_{min}\, M_{max}}$.

The Ratio (β) between the Electrostatic Force and the Gravitational Force

Since Coulomb's law and Newton's law have a similar shape, it is easy to compare the two forces by establishing a ratio (noted β). If we consider the electron-proton interaction (subscript $_{\mathbf{ep}}$), then $\boldsymbol{\beta = k_e\, e^2/G\, m_e\, m_p}$, where m_e is the mass of the electron and m_p is the mass of the proton. This ratio whose value is $\beta \approx 2.268\,7 \times 10^{39}$ was popularized at the beginning of the 20th century by Arthur Eddington and Paul Dirac. Like the fine structure constant (α), this large dimensionless number continues to intrigue.

From the two previous equations, it appears that:

$\boldsymbol{\alpha/\pi = \beta\, m_e\, m_p/M_{min}\, M_{max}}$,

which could contribute to a better understanding of these mysterious ratios.

The Greek Equation (dimensionless)

At this stage, three dimensionless constants are available:

- The max/min ratio: $\Sigma \approx 5.026\,8 \times 10^{120}$
- The fine structure constant: $\alpha \approx 7.297\,353 \times 10^{-3}$
- The ratio between the electrostatic force and the gravitational force: $\beta \approx 2.268\,7 \times 10^{39}$

Numerically, we can see that these values are linked together by the relation $\boldsymbol{\alpha/\pi = \beta^3/\Sigma}$ that I call the Greek equation because of the Greek characters that compose it. Here is why:

The integration of the β ratio to the cosmospheric model implies the existence of a quantity of electrostatic force (or electron-proton) $F_{ep} = \beta\, F_{min}$. Since electrostatic and gravitational forces are conservative, we can associate to the electron-proton pair a quantity of electrostatic energy $E_{ep} = \beta\, E_{min}$. Now, this value of E_{ep} can be calculated precisely using the relation $\boldsymbol{E_{ep}^2 = m_e\, m_p\, c^4}$.

Indeed, this equation is like the one governing the intermediate energy: $\boldsymbol{E_{int}^2 = M_{min}\, M_{max}\, c^4}$. Using the values of CODATA 2018, we calculate: $E_{ep} = \sqrt{m_e\, m_p}\; c^2 \approx 3.508\,2 \times 10^{-12}$ J.

Since $\alpha/\pi = \beta\, m_e\, m_p/M_{min}\, M_{max}$, by multiplying the second member of this equation by c^4/c^4, we get $\alpha/\pi = \beta\, E_{ep}^2/E_{int}^2$.

By replacing E_{ep}^2 by $(\beta^2\, E_{min}^2)$, we get:

$\alpha/\pi = \beta\,(\beta^2\, E_{min}^2)/E_{int}^2 = \beta^3\, E_{min}^2/E_{int}^2$.

Now, $E_{min}^2/E_{int}^2 = E_{min}^2/E_{min}\, E_{max} = E_{min}/E_{max} = 1/\Sigma$, checking the Greek equation: $\boldsymbol{\alpha/\pi = \beta^3/\Sigma}$, which can also be written $\boldsymbol{\Sigma = \pi\, \beta^3/\alpha}$.

Calculating the Expansion Rate (H) and the Age of the Universe

As $E_{min} = h\,H = E_{ep}/\beta$, we deduce that:

$H = E_{ep}/\beta\,h \approx 2.333\ 8 \times 10^{-18}$ s-1,
or **72.013 km/s per megaparsec**.

The age of the Universe would then be:

$H^{-1} \approx$ 13.578 Ga (gigayears)

The precision of these calculations depends on the accuracy with which the value of the physical constants can be determined, mostly that of G and, to a lesser extent, those of k_e, m_e and m_p.

Overall, these values are compatible with observation. The age of ancient stars of very low metallicity, such as «SMSS J160540.18-144323.1» or «2MASS J18082002-5104378 B» is estimated at 13.5 Ga (gigayears). As several ancient stars orbit in the galactic plane, our Milky Way would be very old itself. The most distant galaxy observed to date is GN-z11, as it was 13.4 Ga (gigayears) ago.

In a Big Bang model, the formation of the first stars requires a gestation duration of the Universe, now estimated at several hundred million years.

On the other hand, according to the cosmospheric model, the age of the Universe was already 13.578 Ga (gigayears) at the birth of the first stars. They then would have had all the time required to form.

Scope and Implications of the Greek Hypothesis

The Greek hypothesis implies that the physical constants are related to each other. By developing the equation $H = E_{ep}/\beta\, h$, we get:

$$\boldsymbol{H = G\,(m_e\, m_p)^{3/2}\, c^2 /\, h\, k_e\, e^2}\ \mathrm{s}^{-1}.$$

The age of the Universe would therefore be:

$$\boldsymbol{h\, k_e\, e^2 /\, G\,(m_e\, m_p)^{3/2}\, c^2}\ \text{seconds.}$$

By construction, the dimensionless constants are also related to each other:

$$\boldsymbol{m_e\, m_p / M_{min}\, M_{max} = \beta^2/\Sigma = \alpha/\beta\pi}$$

Likewise, if you wonder why: $\boldsymbol{m_p/m_e \approx 1\,836.153}$
it would be because:

$$\boldsymbol{\sqrt{m_p/m_e} = \beta\, M_{min}/m_e = m_p/\beta\, M_{min} \approx 42.850}$$

In the book «*The Hitchhiker's Guide to the Galaxy*» by Douglas Adams, an advanced civilization wonders about life, the Universe and everything. After 7.5 million years of reflection, the supercomputer «*Deep Thought*» finally provides the answer: 42, which is close to the previous result. The problem is that nobody really knows the question to this answer!

More seriously, the Greek hypothesis is refutable according to the philosophy of Karl Popper. It may or may not be refuted by more precise measurements of *H* or other observations. In any case, this does not call into question the cosmospheric principle, but only the accuracy of the model.

COSMOSPHERIC INTERPRETATION OF MOVEMENT

According to the cosmospheric system, any movement of a massive body relative to another is aimed at maintaining the uniqueness and spatial isotropy of the Cosmosphere. Under the effect of a vectorial force, a massive body (or object) contracts lengths in the direction of this force and dilates them in the opposite direction. This spatial deformation is then felt by the object in the form of an acceleration. With respect to its initial position, the object then moves in the direction of the contracted lengths, so as to remain centered with respect to the Cosmosphere. Indeed, the lengths being no longer the same as in the initial frame of reference, without this displacement, the perceived expansion of the Cosmosphere would no longer be isotropic for this object.

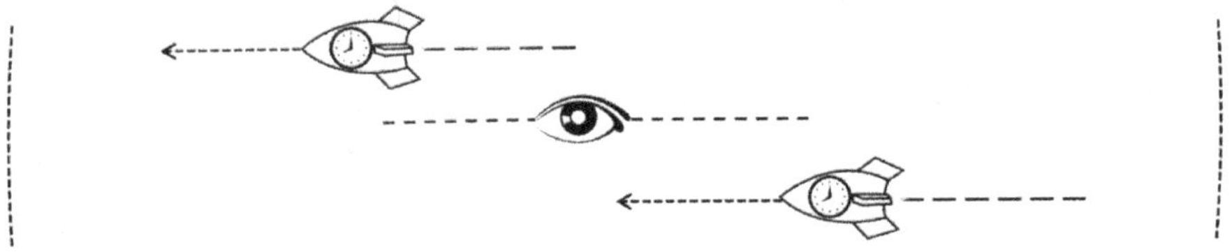

Figure 6 – Spatial deformation for a moving body

In this fanciful schema, the dashes of the central line represent the units of length of the stationary observer, here represented by an eye. From the point of view of this observer, the unit of length of a traveler seems to him contracted when the traveler approaches, whereas it seems to him dilated when the traveler moves away.

These spatial corrections are required so that the speed at which the Cosmosphere moves away from any observer (c) is the same in all frames of reference and in all directions. This mechanism makes it possible to establish a relative speed between objects, but this speed is limited by the maximum speed (c).

The Cosmospheric Interpretation of the Doppler Factor (δ)

The contraction or expansion factor observed is defined by the so-called Doppler-Fizeau effect. The Doppler factor (noted δ) is defined as the ratio between the frequency of the light-emitting source (f_s) and that of the observer (f_o). As the value of c is invariant, this ratio is the same as the ratio between the observed wavelength (λ_o) and that of the emitting source (λ_s). We observe that:

$\delta = f_s/f_o = \lambda_o/\lambda_s = \sqrt{1+\beta}\ /\ \sqrt{1-\beta}$, where $\beta = \pm v/c$

The cosmospheric interpretation of this factor differs, however, from the relativistic interpretation. The latter implies a complex distortion of a hypothetical objective reality. The cosmospheric interpretation simply refers to the subjective reality of the observer.

From the observer's point of view, when an object approaches him, it moves away from the Cosmosphere, and conversely, when it moves away from him, it moves towards the Cosmosphere. This is why one observes an inversion of the Doppler factor when an object in movement crosses the observer.

From the cosmospheric point of view, the factor of temporal contraction or expansion is always the same as that of spatial deformation. Indeed, this is what the relation $L = c\,T$ implies. The age of the Universe always corresponds to the light distance that separates us from the Cosmosphere.

Conceptual Changes - We Cannot Keep Everything

In any conceptual approach, one must identify what is useful for understanding (here, that of phenomena) and prune what is not. Indeed, if we simply introduce new concepts and principles, what becomes superfluous is likely to conflict with what is new. For example, at the beginning of the 20th century, the designers of relativistic equations, Hendrix Lorentz and Henri Poincaré, were still juggling with the notion of ether. Unfortunately, this concept of ether is hardly compatible with Einstein's special relativity. It is often more difficult to abandon certain well-established notions than to introduce new ones. A similar dilemma arises with respect to space-time and the Cosmosphere. According to my analysis, these two concepts are mutually exclusive and a choice must be made:

- the Copernican principle and the concept of objective space-time associated with it, or
- the cosmospheric principle based on the concept of Cosmosphere.

We can consider that space-time is to the cosmospheric principle what ether is to relativity: an obstacle to understanding. Just as the concept of space-time makes the concept of ether superfluous, the cosmospheric principle suggests replacing space-time by the Cosmosphere.

To justify this dissenting position, one must imagine an experiment that can be explained using the concept of the Cosmosphere, where that of space-time fails. To do so, I rely on the previous work of specialists in the field.

The Paradox of the Langevin Twins

The internal coherence of special relativity was quickly questioned. As early as 1911, the physicist Paul Langevin presented a thought experiment that seems to show that special relativity is contradictory. This thought experiment evolved into what is now called the Langevin's twin paradox:

One of the twins remains on earth, while his brother travels in space at very high speed. According to special relativity, the clock of the traveling twin must slow down relative to that of his brother. Upon his return, the traveling twin would be younger than his brother. However, if there is no preferred frame of reference, it can be considered that it is the twin who remains on earth that moves in relation to his brother. It would be his own clock that slows down in relation to his brother's, and he would therefore be the younger one. There seems to be a contradiction.

Opinions on this paradox are divided. Some do not accept the fact that both clocks can slow down, one in relation to the other. If the traveling twin's clock slows down in relation to his brother's, they infer that the brother's clock must speed up in relation to the traveling twin's clock. Despite appearances, this concept of symmetrical slowing down of clocks is perfectly coherent. It can be compared to the perspective effect. When a person moves away from us, his size seems to decrease. This person perceives, however, that it is our height that decreases.

Many believe that the traveling twin will be the youngest. The reason is that the situation is not symmetrical. The twin who stays on earth remains in the same frame of reference throughout the entire experience, while his brother is accelerated several times: once on the outward journey, a second time when he turns around, and possibly a third time to slow down when he returns to earth. These accelerations have the effect of modifying his frame of reference, and thus slowing down his clock in relation to the terrestrial frame of reference.

The Experience of Symmetrical Twins

There is a way to test whether the hypothesis of the asymmetry of the Langevin twins is conclusive. It is a matter of modifying this thought experiment so as to eliminate the asymmetry between the twins:

They first go to the same place, located beyond the Earth's orbit. They are assumed to have zero speed relative to the sun, so that they can fall vertically towards the sun. They are then in a non-accelerated environment, i.e. they do not feel any acceleration. After synchronizing their clocks, the twins are accelerated symmetrically, but in opposite directions, in circular orbits of about one astronomical unit around the sun. These orbits are adjusted so that the period of translation of the twins is one Earth year. The gravitational influence of the planets and moons, including the earth-moon system, is not taken into account here. To avoid collision during the encounter, the orbital planes of the twins are slightly tilted with respect to each other. The twins are then in distinct but non-accelerated environments, a bit like in an orbital station, but around the sun instead of the earth.

The twins cross each other twice a year, at the summer and winter solstices. They then have a relative speed of 60 km/s, one to the other. At the equinoxes, their relative speed is zero. To end the experiment, the twins follow a procedure like that of their departure, decelerating to a meeting point beyond the Earth's orbit.

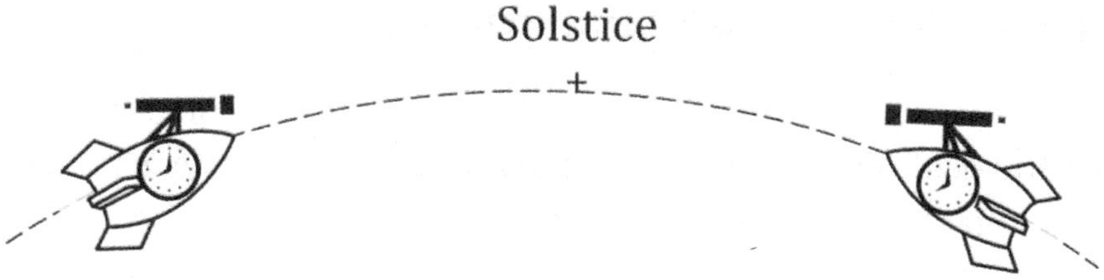

Figure 7 - The experience of symmetrical twins

As the situation is symmetrical, a twin's frame of reference cannot be privileged over that of his brother. After ten circular orbits, both twins will be ten years older and will be of the same age. However, it is known that when the twins' clocks cross each other, they become out of sync because of their relative speed. Special relativity predicts that the clocks should slow down symmetrically with respect to each other.

For the twins to be of the same age when they return, the clocks must resynchronize at least twice a year. Indeed, the experiment can be terminated each time the twins' paths cross. To compensate for the symmetrical slowing down of the clocks, the clocks must therefore accelerate in relation to each other. Time must not only be able to dilate, but also to contract. Unfortunately, special relativity does not allow for time contraction. Similarly, special relativity predicts a symmetrical contraction of lengths when twins cross each other. However, the paths traveled by the two twins are equivalent because of the symmetry of the experiment. To compensate for this symmetrical contraction of lengths, these must be able to dilate too, which is not allowed by special relativity.

As for the gravitational field in which the twins are immersed, it is symmetrical and practically constant during the experiment. It cannot, therefore, be responsible for the half-yearly resynchronization of the clocks. I conclude that Langevin's paradox remains, even in the absence of asymmetry.

The Four Seasons of Symmetrical Twins

The experience of symmetrical twins raises a new question: at what times of the year do clocks slow down relative to each other and at what times do they speed up?

In comparison with the perspective effect, we saw that when a person moves away from us, his or her size seems to decrease, and that this perception is reciprocal. Conversely, when this same person returns towards us, his or her height seems to increase, and vice versa.

In the experience of symmetrical twins, a similar phenomenon is observed: between each crossing, the twins actually move away from each other for three months, either in summer or in winter. They then move towards each other for another three months, either in autumn or in spring. It can therefore be assumed that it is the fact of moving away from or towards each other that determines whether time and space dilate or contract. Now, this is exactly what the cosmospheric interpretation of the Doppler-Fizeau effect confirms.

Let's consider each twin's point of view (their subjective reality). It is assumed that each of them has a big clock and a strong telescope that allows each one to observe his brother and his clock. They also have a real-time audiovisual telecommunication system, i.e. without any buffering delay when transmitting or receiving sound and image.

As the twins move away from each other, light and radio waves take longer and longer to travel the increasing distance between them. The twins therefore perceive that their brother moves and speaks more slowly than normal. Both with their telescope and their video system, they observe that the hand of their brother's clock moves slower than their own, and as a result, it lags. The frequency of the radio wave that carries the sound and image is lower than normal ($f_o < f_s$), and the length of this wave is correspondingly longer ($\lambda_o > \lambda_s$). Twins perceive a dilation of the lengths.

Conversely, as the twins get closer to each other, light and radio waves take less and less time to travel the decreasing distance between them. As a result, the twins perceive that their brother is moving and talking faster than normal. Both with their telescope and their video system, they observe that their brother's clock hand moves faster than their own, compensating for the delay while moving away. The frequency of the radio wave that carries the sound and image is higher than normal ($f_o > f_s$), and the length of this wave is reduced accordingly ($\lambda_o < \lambda_s$). Twins perceive a contraction in length.

Criticism of Special Relativity

According to special relativity, clocks should always be slowed down relative to each other, both when approaching and moving away. Similarly, lengths should always be contracted. This results from the use of the Lorentz factor ($\gamma = 1/\sqrt{1\text{-}v^2/c^2}$) which is always positive, regardless of the directions of the velocity vectors **v** and **c**.

On one hand, this is not what the twins observe during their journey, and on the other hand, these theoretical predictions are incompatible with the measurements made after the experiment.

Thanks to Einstein, we know that the energy of a photon is $E = h\,f$, where f is the measured frequency of the photon. For a photon of frequency (f_S) according to the clock of the emitting source, the observed frequency (f_O) depends only on the relative rate of the clock used to make the measurement. We cannot, therefore, change the observed frequency of photons without also changing the relative rate of the clocks involved. This would be equivalent to changing the intrinsic energy of photons, thus derogating from the principle of energy conservation. Now, this is precisely what special relativity suggests when the twins cross! (?_?)

If $f_O < f_S$, it is because the observer's clock beats faster than that of the emitting source. Conversely, if $f_O > f_S$, it is that the observer's clock beats more slowly than that of the emitting source.

Since its formulation in 1905, special relativity has constantly been criticized from all sides for its lack of coherence. This brief analysis aims to convince you that some of these criticisms are well founded.

The Lorentz Factor vs. the Doppler Factor

From the above, we conclude that the factor to be used to calculate the dilation or contraction of lengths and durations cannot be the Lorentz factor (noted γ), since it is always positive. On the other hand, the Doppler factor (noted δ) seems to have the required characteristics. Let $\beta = \pm v/c$, we can write:

$\delta = (1+\beta)^{1/2}\ (1-\beta)^{-1/2}$, and

$\gamma = (1+\beta)^{-1/2}\ (1-\beta)^{-1/2}$

For a positive value of β, the Doppler factor (δ) is always greater than the Lorentz factor (γ). For values of v close to that of light, the ratio δ/γ then tends towards 2.

There are three reasons for the choice of δ:

- This is what we observe: the ratio $f_S/f_O = \lambda_O/\lambda_S = \delta$.
- The cosmospheric principle requires that the factor used to be the same for durations and lengths, and that it be inverted when two observers cross.
- The principle of conservation of energy requires that the ratio (f_S/f_O) equals the time dilation or contraction factor (δ).

The use of the Lorentz factor by special relativity seems to stem from the concept of inertial (or Galilean) reference frame, which itself stems from the Copernican principle.

According to the cosmospheric principle, the concept of inertial reference frame is a mental construction, a Cartesian idealization that does not exist in nature. This concept can be useful when distances, durations or velocities are negligible compared to the maximum values of these quantities. However, it is precisely outside these conditions that relativistic or cosmological effects occur.

The Concept of Cosmospheric Reference Frame

A reference frame is a reference system that has three space coordinates and one time coordinate. These coordinates are used to define the notions of position, velocity and acceleration. A reference frame is therefore by definition subjective, that is to say that it depends on the observer who makes the measurements.

A cosmospheric reference frame is a reference frame that obeys the cosmospheric principle, which is not the case for the inertial reference frame. This is why the concept of inertial reference frame must be replaced by that of cosmospheric reference frame.

Figure 6 "Spatial deformation for a moving body" partly illustrates the cosmospheric reference frame of a moving observer as seen by a fixed observer.

Because the Doppler factor (δ) is consistent with observation and compatible with the cosmospheric principle, I assume that it correctly translates the reference frame R' of a moving observer (along the x-axis) with respect to that of a fixed observer R.

$\Delta x' = \delta^{\pm 1}\, \Delta x$

$\Delta t' = \delta^{\pm 1}\, \Delta t$

$\Delta x = \delta^{\pm 1}\, \Delta x'$

$\Delta t = \delta^{\pm 1}\, \Delta t'$

If the two observers cross, the intervals Δx and Δt must be decomposed, and the exponent of δ must be inverted. We use (δ^{-1}) before the crossing and (δ^{+1}) after the crossing. The value of β is here considered positive while approaching, i.e., when the vectors ***v*** and ***c*** have the same direction, and thus the same sign.

The Cosmosphere vs. Space-time

Space-time is seen as the backdrop where events occur. This backdrop can be rigid, as Newton assumed, or flexible, as general relativity suggests. In sum, it seems obvious that the physical universe needs a framework that supports its components, and that allows the observer to evaluate distances and duration of events.

The problem is that the experience of symmetrical twins makes the concept of space-time inconsistent. When the twins cross each other, their brother's clock suddenly changes from maximum temporal contraction to maximum dilation. The same phenomenon is observed regarding lengths. These abrupt changes are simply incompatible with the existence of an objective space-time continuum.

To solve this problem, a major conceptual change is required. The framework of the Universe would no longer take the form of an endoskeleton like space-time, but rather that of an exoskeleton: The Cosmosphere, this rigid shell that houses the Universe and allows any observer to evaluate distances and duration of events. Since the concept of space-time is no longer required, the approach used requires its pure and simple abandonment.

I am aware that physical realism is a widespread belief. If you subscribe to it, consider that to assert the physical existence of space-time is to confuse the model with reality, the map with the territory. A concept is supposed to represent reality, not to replace it.

Extending the Life of the Muon

To illustrate the cosmospheric interpretation of time deformation, I will use a well-known observation: the extension of the lifetime of muons in the Earth's atmosphere.

When cosmic rays enter the upper atmosphere, they collide with gas nuclei. This results in unstable particles, which rapidly decay into other particles, including the muon, which is similar to the electron, but much heavier. Because many reach the ground, we know that the lifetime of a muon that passes through the atmosphere at very high speed seems to be prolonged.

A common interpretation of this phenomenon is that if we could observe the muon clock, it would seem to us slowed down compared to the earth clock. This interpretation does not consider the subjective reality of the muon. It does not correspond either, to the analysis made previously.

According to the cosmospheric interpretation, if the muon could observe the Earth's clock, it would seem to him accelerated and vice versa, since this phenomenon is symmetrical. The muon would find that during his crossing of the atmosphere, much more time passes on earth than for him. With the lengths being contracted, the Earth's atmosphere is much thinner for him. This is why he crosses it in a very short time. It is his subjective reality that determines his life span, not ours.

SOME COMPELLING OBSERVATIONS

Several physical phenomena can be explained, at least in part, using the cosmospheric principle. This is applicable on a human scale as in the case of inertia and the centrifugal effect. On a larger scale, the cosmospheric principle can account for universal gravitation and the expansion of the Universe.

Inertia and the Centrifugal Effect

In a circular and uniform motion, the normal component of velocity is constant. However, because of the continuous change in the direction of the velocity vector, this motion requires the presence of a centripetal acceleration, directed towards the center of the circle.

The centrifugal effect is defined as a fictitious force that opposes the centripetal force associated with this acceleration. For example, in a curve, we feel this centrifugal effect that tends to push us out of the curve. Because the earth rotates on itself, the centrifugal effect opposes the earth's gravity. We are therefore slightly lighter at the equator than at the poles. Conversely, we can use this effect to simulate gravity in an environment without it, or to spin our laundry.

The speed of an object is relative and is not felt, but this is not the case for acceleration. When Newton spins a bucket of water, he finds that as the water rotates with the bucket, the surface of the water takes on a concave shape. He then wonders what the water is spinning against. Similarly, we know that a gyroscope maintains its axis of rotation, regardless of the changes in the orientation it undergoes. How can a rotating object «find» that it is rotating, «know» its axis of rotation and «determine» its angular speed ?

Inertia is defined as the tendency of a body to maintain its velocity, both in norm and direction. According to this definition, it is inertia that is responsible for the centrifugal effect. To change the direction of an object, a force (e.g. a centripetal force) must be applied to it. The question is therefore to identify the cause of inertia. To explain inertia and the centrifugal effect, an absolute spatial frame of reference is required, because it is this frame that determines the direction of a moving body, and thus a possible change in that direction.

Mach's principle supposes that the inertia of a massive body is induced by all the other masses of the Universe. To date, we have no idea how these other distant masses could determine the direction and speed of an object. This is an unsolved problem in modern physics.

According to the cosmospheric principle, this absolute landmark is the Cosmosphere. To change the velocity of an object, whether in norm or direction, an adjustment of the lengths is required, as in its setting in motion described above. In a rotating object, each particle of matter must constantly remain centered with respect to the Cosmosphere. To do so, it must therefore continuously adjust its length measurement in the plane of this rotation.

As with the movement of an object, this spatial correction is felt in the form of an acceleration. To change the direction of a translational particle, it must undergo an acceleration (a) directed towards the center of the circle, and whose norm is proportional to the radius of translation (R). This is why we observe that ($a = -\omega^2 R$), where ω is the angular velocity. The negative sign (-) indicates that the centripetal acceleration is directed towards the center of the circle, in the opposite direction of the centrifugal effect felt.

Inertia and the centrifugal effect are tangible proof of the existence of the Cosmosphere and the principle associated with it.

Gravity and Spatial Evolution of the Universe

The cosmospheric interpretation of universal gravitation differs significantly from that of Einstein's general relativity. Indeed, by eliminating the concept of space-time, we can no longer invoke a curvature of this fictitious entity to explain the phenomenon. There must therefore be another explanation based on the concept of Cosmosphere, and which predicts similar effects.

According to the cosmospheric system, gravitation is a dynamic process associated with the spatial evolution of the Universe. To describe this evolution, the use of differential calculus is required. The following equations therefore include derivatives with respect to time, and the Newtonian notation is used. These equations assume the use of matter instrumentation, the one normally used. Here we consider the rate of change H over an infinitesimal duration. As the light time is dilating with respect to that of the matter clock, we calculate:

$\dot{H} = H^2$; $\ddot{H} = H^3$; $\dddot{H} = H^4$; ...etc.

Let $R_U(t)$ be a function that calculates the radius of the Universe at time t, I note:

$\dot{R}_U = H\,R_U = c$, the speed at which the Cosmosphere is expanding, and

$\ddot{R}_U = H\,\dot{R}_U = H^2\,R_U = H\,c$, the cosmic acceleration.

Similar reasoning applies to the volume of space contained by the Cosmosphere.

Let $V_U(t)$ be a function that calculates the volume of the Universe at time t: $V_U = (4/3)\pi\,R_U^3$.

As the expansion operates in each of the three spatial dimensions, we calculate:

$\dot{V}_U = 3H\, V_U = 3H\,(4/3)\pi\, R_U^3 = 4\pi\, H\, R_U^3 = 4\pi\, c^3/H^2$
$\approx 6.216\ 6 \times 10^{61}$ m³/s, the cosmic flow rate, and

$\ddot{V}_U = H\,\dot{V}_U = \mathbf{3H^2\, V_U} = 4\pi\, H^2\, R_U^3 = 4\pi\, c^3/H \approx 1.450\ 8 \times 10^{44}\ \text{m}^3/\text{s}^2$, the variation in the cosmic flow rate.

Now, as we saw earlier, $M_U/V_U = 3H^2/8\pi\, G$ (Friedman's Universe of Critical Density), hence:

$\boldsymbol{\ddot{V}_U = 3H^2\, V_U = 8\pi\, G\, M_U}$

This invariant relationship between V_U and M_U may be the most significant of the cosmospheric model. It shows that the variation in the cosmic flow rate associated with the acceleration of the expansion, and the gravity associated with the mass of the Universe **describe the same phenomenon.**

According to conventional models, there is no reason to observe this relationship. Until recently (1998), we were convinced that the expansion of the Universe was slowing down because of the gravity generated by the mass of the Universe. The acceleration of the cosmic expansion is not only a surprise; its cause is unknown as well as why we observe this value.

From the above, we understand that massive bodies literally create space around them, generating the overall cosmic expansion and flow rate. Thus, we pose $\boldsymbol{M \propto L^3\, T^{-1}}$.

To put things in perspective, let's consider a mass of one kilogram. This mass generates a space flow rate of:

$3H\, V_U/M_U = 8\pi\, G/H \approx 7.187\ 6 \times 10^{8}\ \text{m}^3/\text{s}$.

The increase in this spatial flow rate corresponds to the gravity induced by the increase in this mass of one kilogram:

$3H^2\, V_U/M_U = 8\pi\, G \approx 1.677\ 4 \times 10^{-9}\ \text{m}^3/\text{s}^2$.

To remain centered with respect to the Cosmosphere, massive bodies must constantly adjust their spatial position in relation to each other. In other words, they must obey the cosmospheric principle:

- The inertia of a massive body results from the isotropy of the cosmospheric expansion, because the acceleration of a body requires an adjustment of lengths. Now, this adjustment of lengths influences the sphericity of the space generated by this body. The more massive a body is, the more space it generates around it. This is why the force required to move a body is proportional to its mass.
- The gravity generated by a massive body results from the isotropy of the acceleration of the cosmospheric expansion. To compensate for the increasing creation of space generated by a massive body, the surrounding objects must continuously contract lengths in the direction of the massive body, and dilate them in the opposite direction. This translates into an acceleration of the surrounding objects towards the massive body.

The creation of space around massive bodies is a direct consequence of the high frequency recentering of massive particles with respect to the Cosmosphere, as described above. Under the effect of this incessant bombardment, massive bodies contract, generating a spatial expansion according to the matter ruler.

Since the mass of bodies increases, the net result of this process is an excessive creation of space in the vicinity of the massive bodies, which is interpreted as a curvature of space-time by general relativity. Gravitation is only the cosmospheric correction to this local excess of space. The concept of space-time is not required. As gravitation results from an increasing creation of space, light rays are also deflected.

The Cosmic Expansion Runaway

Since this spatial dynamic results from the very dilation of time, the derivation process continues and the acceleration of cosmic expansion speeds up. For example, third-order derivatives would be:

$\dddot{R}_{\mathrm{U}} = H\,\ddot{R}_{\mathrm{U}} = H^3\,R_{\mathrm{U}} = H^2\,c \approx 1.632\,8 \times 10^{-27}$ m/s³, and

$\dddot{V}_{\mathrm{U}} = H\,\ddot{V}_{\mathrm{U}} = 3H^3\,V_{\mathrm{U}} = 8\pi\,G\,H\,M_{\mathrm{U}} = 4\pi\,c^3 \approx 3.385\,9 \times 10^{26}$ m³/s³.

To compensate for the runaway cosmic expansion, the cosmospheric isotropy requires a constant increase in the gravitational field around massive bodies. On a human scale, the effects of this increase in the gravitational field are tiny. However, in the vicinity of sufficiently massive bodies, they could be detectable.

Dark Energy

The concept of dark energy was introduced to explain the acceleration of cosmic expansion in a relativistic context. From a cosmospheric point of view, this concept is not required. The acceleration of the expansion is only the second derivative of space with respect to time in the same way that expansion is its first derivative. Rather, these observations are observational evidence of the expansion (at rate H) of light time relative to matter time.

The Cosmospheric Interpretation of the Action-Reaction Principle

We have been taught that the gravitational force (F) between two masses (M and m) is $F = G\,M\,m/r^2$, where r is the distance between the two masses.

In fact, Newton's formula exposes only half of the impact resulting from the gravitational interaction between masses M and m. It expresses well, for example, the gravitational component ($F_{M/m}$) that the earth (of mass M) exerts on an apple (of mass m), but it should be noted that this apple exerts on the earth an equivalent but opposite gravitational component ($F_{m/M}$). This is called the action-reaction principle, or Newton's third law of motion. The force of attraction earth/apple ($F_{M/m}$) cannot exist without the force apple/earth ($F_{m/M}$) and vice versa.

The same principle applies to other forces. For instance, the force of attraction that a proton exerts on an electron ($F_{\mathrm{p/e}}$) is balanced out by the force that the electron exerts on the proton ($F_{\mathrm{e/p}}$).

On the one hand, we have seen that a force exerted on a massive body contracts the units of length in one direction and dilates them in the opposite direction. On the other hand, the mass of a body would be proportional to the space flow it generates, thus contributing to the cosmic expansion.

An equivalent force, but in the opposite direction, is therefore required to respect the isotropy of the cosmospheric expansion. This is the cosmospheric explanation of the action-reaction principle.

The Recession of Distant Galaxies

To share the Cosmosphere with us, a distant object must have a unit of length dilated in our direction, and contracted in the opposite direction. From its point of view, the distant object then remains centered with respect to the Cosmosphere. As in the cosmospheric interpretation of motion, this object then moves in the direction of the contracted lengths, so as to respect the isotropy of the cosmospheric expansion. In other words, it is moving away from us at a speed that increases with distance (Hubble's law).

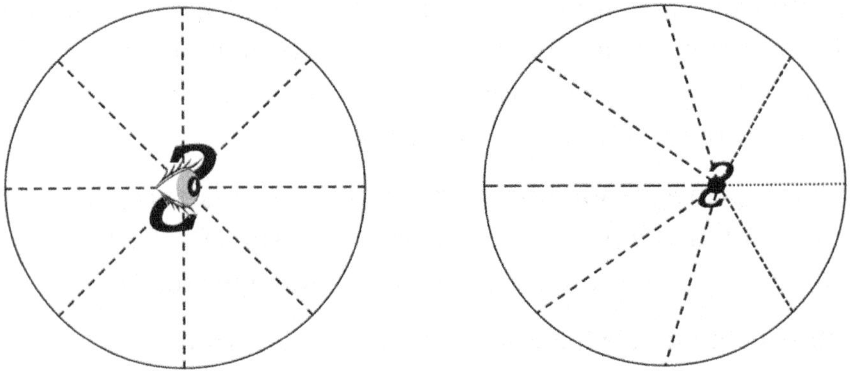

Centered galaxy **Distant galaxy**

Figure 8 - The recession of distant galaxies

On the left figure, the dashes represent the units of length for the observer of the centered galaxy. On the right figure, the dashes represent the units of length of the distant galaxy, as perceived by the observer of the centered galaxy. This spatially symmetrical phenomenon alone explains the recession of distant galaxies. It is thus the same principle that governs the movement of near and distant objects.

It is important to note that this recession of distant galaxies implies the use of a matter instrumentation. According to the light ruler, the distance that separates us from a galaxy (e.g. 0.1 R_U) always remains the same, if we disregard other forces and the proper motion of this galaxy with respect to ours.

The Relative Position and Speed of Objects

The previous reasoning on the position and recession of galaxies applies in principle to all objects in the Universe. Indeed, as all objects remain centered with respect to the Cosmosphere, and none of them occupy the same location, this implies that they all have different units of length. Consequently, two distinct objects cannot be totally motionless with respect to each other. The relative speed of an object with respect to another must be at least v_{min}. As this value is tiny, this phenomenon cannot be observed in practice.

THE TEST: GRAVITY VS. COSMIC EXPANSION

This section tries to identify some observations that would allow distinguishing between the cosmospheric hypothesis and that of conventional theories.

The basic idea is that the cosmic expansion associated with the cosmospheric system differs from that predicted by the Big Bang theory. One observable difference concerns the distance at which gravitation gives way to cosmic expansion. Some astronomical observations could therefore favor one interpretation to the detriment of the other.

The Conventional Hypothesis

According to cosmological models such as ΛCDM, the expansion of the Universe is not a force, but rather a rate, the rate at which space itself expands. This spatial dilation would only generate a perceptible pseudo-speed when no attractive force can counter its cumulative effect. For gravity to be able to counter spatial expansion at a distance D from a mass M, the escape velocity associated with this mass M must be greater than the pseudo-speed of the expansion calculated at this same distance D.

For comparison, let's calculate the distance D_Λ at which a body could theoretically escape the gravitational pull of the sun due to spatial expansion according to this ΛCDM hypothesis.

The solar mass is noted $M_\odot \approx 1.988\ 4 \times 10^{30}$ kg.

The expansion pseudo-speed at this distance would be $H\,D_\Lambda$ and the escape velocity, $\sqrt{2G\,M_\odot/D_\Lambda}$. So we have the relation: $H\,D_\Lambda = \sqrt{2G\,M_\odot/D_\Lambda}$, hence $D_\Lambda = \sqrt[3]{2G\,M_\odot/H^2} \approx$ **386 ly**

The pseudo-speed of the expansion $H\,D_\Lambda$ at this distance would be the same as the escape velocity, about 8.5 m/s.

In fact, this distance D_Λ is very large, and the volume of space bounded by this radius contains many stars gravitationally linked together. Spatial expansion would therefore be inoperative at this scale. According to this hypothesis, spatial expansion would only be perceptible at a scale larger than that of galaxy clusters!

The Cosmospheric Hypothesis

On the other hand, the cosmospheric hypothesis assumes that cosmic expansion is a force also present at a smaller scale. To calculate the distance $D_\odot$ from which a light object would escape the solar gravity, the cosmic acceleration $H\,c$ should be equal to the gravitational acceleration of the sun at this distance, i.e. $G\,M_\odot/D_\odot^2$. We calculate:

$D_\odot = \sqrt{G\,M_\odot/H\,c} \approx 4.355 \times 10^{14} \approx$ **2 911 au** or **0.046 ly**

We now have two different predictions for the distance at which solar gravitation gives way to cosmic expansion. By checking whether an object located more than 2 911 au away from the sun is subject to cosmic expansion or not, one might therefore favor one hypothesis to the detriment of the other.

The Oort Cloud

Thanks to the observation of so-called long-period comets, the existence of the Oort cloud is inferred. It is a halo of matter that counts more than 10^{12} comets, located between 20 000 au and 100 000 au of the sun. It is known that comets formed much closer, at an estimated distance of between 50 au and 100 au from the sun. It is hard to understand how the distance of comets could have increased by three orders of magnitude to form a halo of 3 ly in diameter!

If we assume that the recession speed of comets is the product of distance multiplied by the expansion rate, then a comet located at 100 000 au of the sun would be moving away from us at a speed of 1 100 km/a. Considering the cosmic expansion over a duration of 4.6 Ga (gigayears), the cumulative distance is measured in tens of thousands of au.

This observation supports the cosmospheric hypothesis that the comets in the Oort cloud are indeed subject to cosmic expansion. Without a similar mechanism, it is difficult to account for the estimated number of comets in the Oort Cloud.

THE EXPANSION OF THE SOLAR SYSTEM

Some observations suggest that the distance at which cosmic expansion operates is not clear-cut. This force would combine with the force of gravitation, and its effect would be perceptible below $D_{\odot} \approx 2\,911$ au.

The Kuiper Belt

The Kuiper Belt is an area of the Solar System extending beyond Neptune's orbit, between 30 au and 55 au. It would contain more than 70 000 bodies more than 100 km in diameter, including the dwarf planets Pluto, Makemake and Haumea. According to current models of solar system formation, the Kuiper Belt should have at least 30 Earth masses. This mass is required for objects larger than 100 km to have formed at this distance. However, the mass inferred by all observed objects is less than 0.1 earth mass. The cosmospheric hypothesis is that the Kuiper belt would have formed closer to the sun, where its density would have allowed the formation of the observed objects. Like for the Oort cloud's comets, the cosmic expansion would be responsible for the observed distance.

Extreme Trans-Neptunian Objects

Among the objects located beyond the Kuiper Belt, some are very far from the Sun and have a strongly eccentric orbit. This is the case of Sedna whose diameter of about 1 000 km makes it a candidate for the status of a dwarf planet. Its orbit oscillates between 76 au and 955 au over a period of 11 700 years. As for the orbit of 541132 Leleākūhonua (nicknamed The Goblin), it oscillates between 65 au and 2 100 au over a period of 36 000 years!

It is hard to understand how these scattered objects can transit on such distant and eccentric orbits. On the one hand, they are too far away to be disturbed by planets, and on the other hand, according to the current theory, the eccentricity of their orbit should decrease, which obviously does not seem to be the case. Some assume the existence of a hypothetical planet 9, of which no trace has yet been observed.

The cosmospheric hypothesis is that due to space expansion, the eccentricity of orbits would be increasing, and not the other way around. This is due to the fact that satellites spend more time near the apoapsis than near the periapsis. The apoapsis is therefore subject to a greater increase than the periapsis, and this phenomenon becomes more pronounced as the eccentricity increases. This phenomenon could also explain the observation of an unexplained increase in the eccentricity of the lunar orbit.

The Migration of Ice Giants

We know today that the planets Uranus and Neptune did not form where we observe them, i.e. on orbits so far from the sun (19 au for Uranus and 30 au for Neptune). These icy giants which have respectively 15 and 17 Earth masses would have formed closer to the sun and would have migrated thereafter on their current orbits. The cosmospheric hypothesis is that for gigayears, these planets have been slowly moving away from the sun due to the expansion of the outer solar system.

The Spiral Orbit of Planets and Other Satellites

For a planet to move away from the sun, an ascending spiral orbit is required. The following figure is not to scale and neglects the eccentricity of the heliocentric orbit. Such an orbit implies a radial component of the satellite's trajectory.

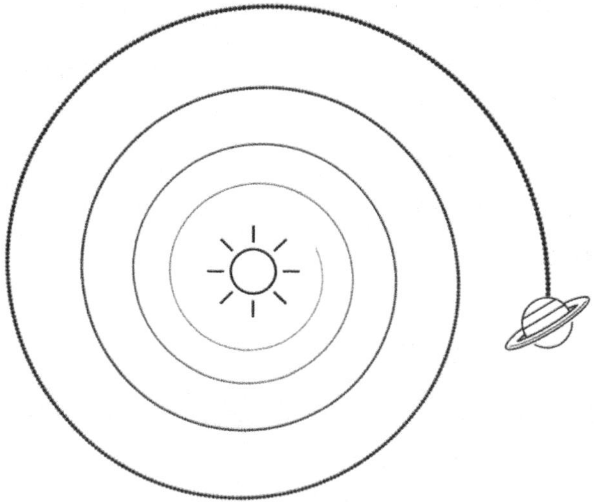

Figure 9 - The spiral orbit of planets and other satellites

This recession of planets results from a decrease (at rate $-H$) in the acceleration vector directed towards the sun. Indeed, if R is the distance between the planet and the sun, the value of this acceleration (a) is provided by the equation: $a = G\,M_{\odot}/R^2$ (or: $a = G\,M_{\odot}\,R^{-2}$). Now the growth rate of $M_{\odot}$ is H, while that of the surface R^2 is $2H$. We calculate that the rate of change in the centripetal acceleration is: $\dot{a}/a = H - 2H = -H$.

Even if the acceleration decreases and the planet moves away, the force connecting the two stars remains constant. Indeed, if $F = m\,a$, then: $\dot{F}/F = \dot{m}/m + \dot{a}/a = H + (-H) = 0$.

The Inner Solar System

At the scale of the inner solar system, solar gravity is largely dominant and spatial expansion would be tiny. However, some observed orbital anomalies could find an explanation of cosmospheric origin:

- An unexplained increase in the earth's orbit (Iorio, 2011).
- An unexplained increase in the eccentricity of the Moon's orbit (Iorio, 2011).
- A planetary flyby anomaly that implies an unexplained acceleration of space probes during gravity assist maneuvers.

The Faint-Young-Sun Paradox

In 1972, astronomers Carl Sagan and George Mullen observed that at the beginning of the Archean period, 4 Ga (gigayears) ago, liquid oceans had long covered a large area of the Earth (Sagan & Mullen, 1972). However, we know that at that time, solar radiation was about 30 % lower than it is now, which implies that the habitable zone was closer to the sun and that the Earth should have been completely frozen. This is called the faint-young-sun paradox.

Since traces of microbial life have been found dating to this remote time, we know that the Earth was not only habitable, but also inhabited and has remained so since. Weirder still, we know that liquid water was also abundant on Mars, which yet orbits between 1.4 au and 1.7 au from the sun.

The conventional assumption is that then, Earth and Mars were covered by a dense atmosphere rich in greenhouse gases, mainly composed of carbon dioxide (CO_2). The problem is that in 2016, the size of gas bubbles trapped in a 2.74-gigayear-old lava flow establishes that the atmospheric pressure on Earth was less than 0.5 atmosphere at that time (Sanjoy M. Som, 2016), disproving this hypothesis. This result is corroborated by nitrogen and argon isotope studies of fluid inclusions in 3.0 to 3.5-gigayear-old rocks (Marty, Zimmermann, Pujol, Burgess, & Philippot, 2013).

The cosmospheric explanation of this paradox is simple. If liquid water was abundant on Earth and Mars at that time, it is because they were orbiting within the habitable zone, i.e. closer to the sun. Due to cosmic expansion, the Earth and Mars would be slowly moving away from the Sun that is warming up, thus remaining in the habitable zone. This slow migration of planets would be required for life to persist and evolve over gigayears, not only on Earth, but also on possible exoplanets harboring life.

THE FORMATION PROCESS OF THE LARGE STRUCTURES OF THE UNIVERSE

The formation and evolution of galaxies form a very active research domain in astrophysics. This domain can be extended to other large structures such as clusters of galaxies and globular clusters. This field is relatively complex and many factors influence the formation and evolution of these structures. However, the proposed method suggests the use of a simplified model that would allow a partial explanation of the origin and evolution of these structures.

Suppose a gigantic cloud of gas and dust floating in intergalactic space. The cosmospheric system assumes the existence of two opposing forces acting on this cloud:

- The cosmic expansion that seeks to disperse matter at the periphery.
- A contraction of the central mass resulting from the gravitational attraction generated by this mass.

The increase in the mass of the particles would contribute significantly to reaching the Jeans mass, i.e. the critical mass required for the collapse of a gas cloud whose overdensity zones are subjected to the pressure of the gas they contain.

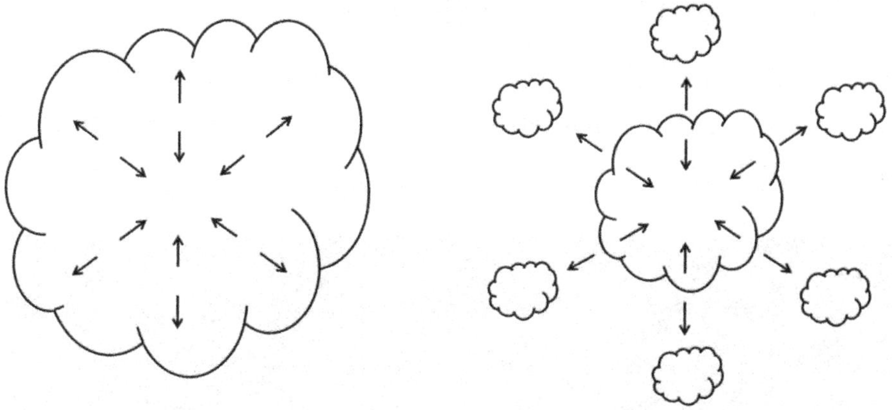

Figure 10 - The formation process of the large structures of the Universe

These two opposing forces tend to fragment the original cloud. A central cloud gains in density and begins a collapse. At the periphery, small clouds form and slowly move away under the effect of expansion. This mechanism would repeat at different scales as the mass of particles increases. It would help trigger the formation of different structures such as:

- A cluster of galaxies that would come from a primitive cloud that would have fragmented.
- A massive galaxy with dwarf galaxies and globular clusters at the periphery.
- A stellar cluster that would come from a cloud of gas and dust that would have fragmented.
- A stellar system with a halo of comets at the periphery.

Observation of Globular Clusters

The observation of globular clusters seems to corroborate the formation mechanism of the large structures of the Universe described above.

Figure 11 - Omega Centauri globular cluster (NGC 5139) Photo by ESO-VST (1119 b)

Omega Centauri is a globular cluster of the Milky Way located about 16 000 ly from the Sun in the constellation Centauri. It is visible to the naked eye and has an angular diameter comparable to that of the moon. It has about 10 million stars, some of which are 12 gigayears old. Its center would contain an intermediate black hole of about 12 000 $_{\odot}$. This formidable structure could be the heart of a dwarf galaxy swallowed by the Milky Way, suggesting a similarity in the formation process of globular clusters and galaxies.

The average distance between the stars at the center of this globular cluster is estimated at only 0.1 ly, about twice the distance $D_{\odot}$ calculated previously. Considering that the average mass of the stars in this cluster is about 0.4 $M_{\odot}$, the cosmic expansion at this scale could explain why this compact mass of stars does not collapse on itself.

According to the cosmospheric hypothesis, globular clusters have been evolving like fireworks, expanding very slowly over gigayears. This would explain the formation and maintenance of their structure, despite the gravity generated by their increasing mass.

Observation of Galaxies

Just like globular clusters, galaxies are very old structures. The resemblance of some galaxies with globular clusters is striking. A globular cluster looks like a model of an elliptical galaxy. Although the scale is different, the mechanism of formation would be similar, and the spatial expansion would be present at the scale of galaxies.

Figure 12 – Elliptical galaxy Messier 7 (NGC 4486) Photo by ESO-VST (1907 b)

The Messier 87 galaxy (abbreviated M87 and cataloged NGC 4486) is a gigantic elliptical galaxy located near the center of the Virgo Cluster, which has about 2 000 galaxies. M87 is about 53 Mly away, and its mass is estimated to be 200 times that of the Milky Way. At its center, there is a supermassive black hole (M87*) of about 6.5×10^9 $M_\odot$. This black hole was imaged in 2019 by the Horizon Telescope Event.

The small dots that we observe at the periphery are not stars. They are mostly globular clusters like the one in the previous figure. M87 has about 12 000 globular clusters, which is huge compared to the 150 contained in the halo of our Milky Way.

The fractal nature of this structure strongly suggests the existence of a formation mechanism like the one described above. The center of a gas cloud would have collapsed to form the heart of the galaxy. At the periphery, the cosmic expansion would then have caused the dispersion of dwarf galaxies and globular clusters that are slowly moving away. Eventually, the stars would have formed according to a similar pattern.

THE ROTATION CURVE OF GALAXIES

When we observe the stars at the periphery of galaxies, we see that they rotate too quickly compared to the distribution of visible mass (stars and gas) in the galaxies. This is called the problem of the rotation curve of galaxies. The same phenomenon is observed in clusters of galaxies. The galaxies rotate too quickly compared to the observed mass. To explain this anomaly, two competing hypotheses are proposed:

- The presence of dark matter halos (or invisible) encompassing the galaxies and clusters of galaxies.
- A modification of the universal gravitation which would be perceptible when the gravitational acceleration is tiny.

Dark Matter Halos

The dark matter halos surrounding galaxies and clusters of galaxies would be invisible, but their presence is inferred from the gravity generated by their mass. According to the formation process previously described, such a halo of matter should indeed exist. However, the invisibility of this matter remains unexplained and its mass undetermined.

The Nature of Dark Matter

The nature of dark matter is unknown for the moment. However, the approach used suggests a hypothesis that I consider as the most likely: dark matter could be made of ordinary matter, but difficult to detect. The more observational techniques improve, the more ordinary matter is observed in the cosmos:

- We detect more and more dark and massive galaxies that emit very little visible light.
- We detect very hot gases in the filaments of matter of the intergalactic medium. The global mass of these gases is largely superior to that of all galaxies.
- Cooler and more tenuous gases could therefore exist. Some of them could be located on the edges of the great cosmic voids, and not only in the galactic filaments.

The Exotic Dark Matter Hypothesis

The concept of exotic dark matter was introduced to try to save the Big Bang hypothesis. On the one hand, the observed abundance of deuterium is incompatible (too low) with a universe composed exclusively of ordinary matter. On the other hand, exotic dark matter is required to explain the formation of the large structures of the Universe such as galaxies and clusters of galaxies. Indeed, if we assume a hot big bang, then the observed fluctuations of the cosmic microwave background are much too weak to explain the initiation of these structures from ordinary matter, because the temperature and radiation pressure were too high.

We then imagined a new type of matter that would not be affected by this radiation pressure: non-baryonic dark matter. This would interact only gravitationally and could therefore have caused the slight density fluctuations observed in the cosmic microwave background. These gravitational potential wells would then have attracted the cooled ordinary matter.

Unfortunately for the Big Bang hypothesis, the observation of this exotic matter is not there. Despite decades of research, no trace of non-baryonic matter has been detected. This is a major problem, because without this exotic substance, the whole big bang theory collapses.

This problem does not exist in the cosmospheric model which assumes a cold primordial universe.

The Theory of the Modified Newtonian Dynamics

The theory of the Modified Newtonian Dynamics (or MOND) was proposed by the astrophysicist Mordehai Milgrom in 1983. It has evolved and been perfected since (Milgrom, 2020). It suggests a modification of Newton's second law at very low accelerations, i.e., below an estimated value $a_0 \approx 1.2 \times 10^{-10}$ m s^{-2}.

In a Newtonian regime, the speed (v) of a star in a circular orbit around a central mass M is a function of the square root of the inverse of the radius R: ($v^2 = G\,M/R$). According to MOND, when the gravitational acceleration (a_g) is very weak ($a_g << a_0$), this speed stops decreasing and depends only on the mass M: ($v^4 = G\,M\,a_0$). This modification explains the rotation curve of galaxies correctly, and that of galaxy clusters partly. Some dark matter is still required in the latter case.

The Observed Value of a_0

The proponents of the MOND theory soon noticed that the value of the acceleration a_0 was of the same order of magnitude as the product of the speed of light (c) and the expansion rate of the Universe (H_0). In fact, we observe that $a_0 \approx H\, c/2\pi$, suggesting a close relationship between gravitation and cosmological parameters.

From a cosmospheric point of view, the magnitude of an acceleration cannot be less than $a_{min} = H\, c$. Therefore, one should forget a formulation like ($a_g << a_0$). We have seen previously that the gravitational attraction of a satellite should gradually decrease, and that its ascendent spiral orbit should move it slowly away from the central mass. Acceleration being a quantified quantity, when it becomes very weak, it should take almost integer values: ... 3 a_{min}, 2 a_{min}, and finally 1 a_{min}.

This minimum value of 1 a_{min} is uncrossable. As with velocity, any object must be accelerated relative to another since these do not occupy the same location. The observed value a_0 is less than that of a_{min} by a factor of 6 ($\approx 2\pi$). I deduce that the vector $\boldsymbol{a}_{min}$ is tilted in the direction of the star's trajectory. The decrease in the radial component of $\mathbf{a}_{min}$, coupled with its tangential component, allows the star to maintain an ascending spiral orbit, while maintaining a constant velocity $v \approx \sqrt[4]{G\, M\, H\, c/2\pi}$. The observation of the rotation curve of galaxies is therefore consistent with the cosmospheric system.

THE PAST OF THE UNIVERSE: A COLD PRIMORDIAL UNIVERSE

The cosmospheric paradigm implies the absence of a hot Big Bang, as well as the omnipresence of cold and light matter, in a very distant past. Cosmological models involving a cold primordial universe face two major objections: one concerns the origin of the cosmic microwave background, and the other, the abundance of light elements observed today. A coherent cosmological model must suggest answers to these objections, while trying to explain the origin of the large structures of the Universe.

The Cosmic Microwave Background

The cosmic microwave background is now sufficiently mapped to allow us to refute certain scenarios or working hypotheses. It is necessary to allow explaining not only its origin, but also its high global isotropy and low local anisotropies.

The global isotropy of the fossil radiation confirms its very ancient origin, and therefore very close to the Cosmosphere, itself perfectly isotropic. It is thus not required to use an inflationary scenario to explain this property.

Regarding the origin of this radiation, the hypothesis of conventional theories seems the most probable. It would be a radiation coming from a very old haze of particles, essentially composed of ionized hydrogen (free protons and electrons) and photons, the whole in a state of thermal equilibrium. These protons and electrons would have combined to form neutral hydrogen, and the haze would then have dissipated. The remaining combination radiation would have reached us in the form of a diffuse microwave background with a characteristic spectrum.

The Origin of the Cosmic Plasma

For the early universe to have been in this plasma state, it is required and sufficient that the kinetic energy of the particles has been higher than their binding energy. In the absence of a hot Big Bang, another mechanism must account for this cosmic plasma. The hypothesis that seems the most probable is that initially the matter was not ionized, but in a neutral state. We can imagine free, cold and light neutrons that spontaneously decay, emitting the ionized particles according to the same process as today. This beta decay is an exothermic reaction that also involves the emission of an electronic antineutrino:

$$\mathbf{n \rightarrow p^+ + e^- + \bar{\nu}_e}$$

The cosmic microwave radiation would have been emitted at the same temperature as observed today (about 2.725 K). The binding energy of the electron in the hydrogen atom (today 13.6 eV) was therefore much lower in the past. This increase in particle binding energy is compatible with the increase in mass described above. The early universe was weakly energetic compared to the one we observe today.

The Abundance of Light Elements

According to the above, hydrogen in the past was very light compared to what it is today. The amount of matter required to form a critical mass and trigger the process of nuclear fusion was much larger then. The first stars were therefore gigantic and would have burned large quantities of hydrogen.

At the end of their life, these mastodons would have exploded, generating cataclysms of galactic scale: the hypernovas. These titanic explosions would have blown away the outer layers of giant stars, dispersing the light elements including helium, lithium and beryllium.

The remains of these mastodons would be at the origin of supermassive black holes that we observe at the heart of galaxies. As the particles' mass was increasing, this process would have repeated itself at different scales, generating the massive and intermediate black holes that we observe today.

The Observed Abundance of Deuterium

Despite the above, the observed abundance of some light elements suggests another mechanism of formation. This is notably the case of deuterium (an isotope of hydrogen) whose origin seems incompatible with stellar nucleosynthesis. Indeed, this element is rapidly burned in stars. However, it is observed almost everywhere, even in intergalactic gas clouds.

On the one hand, it is assumed that at a remote time, the binding energy of particles was very low. On the other hand, it is assumed that at that time, there were free neutrons as well as protons resulting from beta decay. This mixture is indeed the one required for primordial nucleosynthesis. However, an essential ingredient is missing: Heat.

This situation is comparable to that of a gasoline engine. Even if the correct air/fuel mixture is introduced into the cylinder, combustion cannot ignite by itself. An intense heat source is needed to initiate the reaction, such as the spark of a spark plug. Once the exothermic reaction has started, it is self-sustaining due to the heat generated.

First Cosmic Sparks

If the hypothesis of primordial neutrons is correct, then cosmic sparks were indeed generated, and this may well have initiated some nuclear reactions in a low energy context. Indeed, there is another form of neutron decay:

$$\mathbf{n \rightarrow p^{+} + e^{-} + \bar{\nu}_{e} + \gamma}$$

This radiative decay of the neutron occurs about 0.3 % of the time, the other 99.7 % corresponding to the beta decay. The energy of the gamma photons (γ) emitted during this reaction is today between 15 keV and 340 keV, which corresponds to a temperature ranging between 174 MK and 3.9 GK.

All proportion kept, the energy of the photons emitted during the radiative decay of neutrons was thus sufficient to initiate nuclear reactions in a low energy context. Primordial photons have probably been absorbed by matter, and I do not know if some of them could have left any imprint on one of the frequencies of the cosmic background.

A low mass implies a larger cross section of particles and nuclei, which greatly facilitates collisions. We can therefore assume that the origin of deuterium is cosmological in nature:

$$\mathbf{n + p \rightarrow D + \gamma}$$

Indeed, this reaction does not involve any electrostatic repulsion between particles.

Regarding the formation of the other light elements such as helium, lithium and beryllium, it could be of stellar origin as mentioned above. It is true that a low-binding energy suggests a low charge, which would reduce the repulsion between the nuclei by a quadratic factor. However, I do not know what nuclear reactions would be allowed by the conditions present at that remote time.

In summary, the hypothesis of primordial neutrons suggests the occurrence of a primordial nucleosynthesis that would have generated deuterium and possibly some other light elements.

The Origin of the Large Structures of the Universe: The Cryogenesis

Analysis of the fluctuations of fossil radiation shows that primordial megastructures of the Universe have left their imprint on this radiation. We can deduce that the formation of these structures began before the matter radiation decoupling, responsible for the emission of the fossil radiation. We have seen that this poses a problem for hot Big Bang type models, because to initiate the process of gravitational collapse, matter must be stabilized.

On the other hand, if the primordial universe was composed of very cold and ultra-light neutrons, then matter was originally in a different state. Such a fundamental state of matter (that of the lowest energy) was predicted as early as 1925 by Albert Einstein. It is the Bose-Einstein condensate, some forms of which are produced today in the laboratory at very low temperatures.

Under the effect of cosmic expansion, this agglutinated substance would have gradually been transformed into cosmic foam, of which we can still observe the vestiges: large cosmic voids whose diameter is measured in hundreds of millions of light-years. The spongy nature of the Universe at this scale is the signature of a light and agglutinated matter, which would have swollen.

The walls of these cosmic bubbles now form large sheets of galaxies. At the junction of these megastructures, where the density of matter was once greater, we find filaments of matter: the superclusters of galaxies.

Under the combined effect of the cosmic expansion and gravity, this matter would have gradually fragmented, initiating the formation of clusters of galaxies, galaxies and globular clusters, as described above.

During this cryogenic process, neutrons gain mass and energy. At some point, they would have disintegrated into a gigantic cosmic popcorn. The ionized particles would then have combined to form neutral hydrogen. When the Universe became transparent, the remaining cosmic radiation would then have mapped the primordial structures.

The Absence of Magnetic Monopoles

According to the Big Bang theory, there should exist magnetic monopoles, i.e. particles that would carry a point magnetic charge (North or South). However, all observed magnets have two opposite magnetic poles. Where are the magnetic monopoles hidden?

This problem does not exist in the cosmospheric system. The creation of a magnetic monopole theoretically requires a very high temperature. However, according to the cosmospheric hypothesis, such a temperature has never existed.

The Almost Total Absence of Antimatter

First of all, it is good to remember that the existence of antimatter is not only theoretical. Every day, it is produced in small quantities in particle accelerators.

The stars and gas clouds we observe are entirely made of ordinary matter such as protons, neutrons and electrons. According to the Big Bang theory, the observable universe initially contained an equal quantity of matter and antimatter. The temperature of the universe was then very high. Where is the antimatter hiding, or if it no longer exists, how did it disappear?

According to the cosmospheric hypothesis, the observable universe was once composed of cold neutrons that disintegrated. Because time flows from the past to the future, only ordinary matter was able to form and persist in this environment. An antimatter particle is like a matter particle going back in time. It will quickly meet a matter particle and will be annihilated.

Is There a Mirror Universe?

Strangely, the cosmospheric system suggests the existence of a mirror universe. According to the Cosmospheric hypothesis, what is unique and invariant is the Cosmosphere itself. Nothing prevents that a mirror universe, composed of antimatter, was generated by the Cosmosphere, parallel to the universe of matter that we inhabit. In the mirror universe, the arrow of time would be reversed, as well as the left and the right.

The creation of primordial neutrons in the observable universe would have been accompanied by the creation of an equal quantity of antineutrons in the mirror universe. This one would have evolved in a similar way to ours, and would have anti-atoms, anti-stars, anti-planets and maybe even living beings composed of antimatter!

By analogy, let us imagine a world in two dimensions. Let's suppose that we realize that we are located exactly at the center of a coin, on the face side. The circumference of this coin plays here the role of Cosmosphere. It is invariant and constitutes the limit of our space in two dimensions.

Even if we do not have access to it, we can imagine that there is a tail side to this coin, and that this tail side resembles the face side that we inhabit. In this configuration, the head and tail sides share the same circumference. If we try to imagine what is written on the tail side, we realize that the left and right are reversed with respect to the head side.

The hypothesis of a mirror universe solves the problems of spatial and temporal asymmetry. The direction of the arrow of time that we observe would no longer be unique nor arbitrary. The reflection symmetry also allows a left-right inversion of chiral processes and systems.

THE FUTURE OF THE UNIVERSE

Conventional cosmological models predict a sad fate for the Universe. It could be a thermal death of the Universe in a freezing cold, or a torrid melting of matter in a big crunch. These predictions contrast with the optimistic prognosis of the cosmospheric model. According to the latter, the long-term future of the universe is characterized by a slow growth of the mass of particles and stars.

The Critical Mass to Form a Star

For a star to be able to sustain a hydrogen fusion reaction, it must have a minimum mass of about 0.08 $M_\odot$ which corresponds to about 80 times the mass of Jupiter (M_J). Below this critical mass, we are dealing either with a brown dwarf (>13 M_J) or a gas giant (≤13 M_J). These sub-stellar gaseous objects are very abundant and essentially composed of hydrogen and helium, just like stars. They are sometimes called failed stars because they are considered unable to initiate the hydrogen fusion process.

The Magic of the Cosmospheric System

As soon as a concept of critical mass is involved, the increase in mass predicted by the cosmospheric system comes into play. If the mass of particles is growing, then less and less are needed to form the critical mass required for hydrogen fusion. We deduce that the gaseous sub-stellar objects are in fact stars in gestation.

Under the gravitational effect caused by the increase in its own mass, a massive brown dwarf will heat up in its center and eventually ignite to give birth to a real star. This process is slow, but it applies in principle to all gaseous objects whose composition is like that of stars, i.e. rich in hydrogen.

According to the cosmospheric system, Jupiter and Saturn have a bright future ahead of them. In the distant future, they will light up to illuminate their procession of planets, some of which may host astonishing new forms of life.

METAPHYSICAL QUESTIONING

In a systemic approach, a global vision of the system is required. Since the system under study (the Universe) comprises in principle the totality of what exists, epistemological, ontological and theological questions cannot be avoided. This book does not pretend to answer all these questions, but it tries to identify some of the implications of the cosmospheric paradigm.

A priori, the cosmospheric principle postulates the existence of a unique and immutable entity that generates and governs time, space and matter. Now, this immaterial, uncreated and eternal entity is by definition transcendental.

Epistemological Reflections

Consider the following assumptions:

1. The earth is the center of the Universe.
2. The sun is the center of the Universe.
3. The Universe has no center.
4. The center of the Universe is everywhere.

These hypotheses concern physical cosmology. We can, with the help of observation and reasoning, form a more or less coherent image of the Universe, depending on the hypothesis we favor. The choice we make influences directly our interpretation and understanding of phenomena. These four hypotheses seem a priori mutually exclusive and thus irreconcilable.

The question of whether these hypotheses have a truth value is a delicate one. It all depends on what we mean by the Universe. For example, the cosmospheric interpretation of hypothesis 4 refers to a universe to oneself, a projection of the Universe in itself. However, the latter remains inaccessible, and the question of whether it has a center or not is inappropriate, since space would be a physical concept that applies only to a projection, not to the Universe per se.

Now, if we consider this universe to oneself, then hypothesis 4 seems compatible with the first two, namely that we can just as well consider the earth or the sun as the center, since it is everywhere. On the other hand, the Universe would not really have a single, defined center, which is similar to hypothesis 3. As a manifestation of the Absolute Maximum, the Cosmosphere seems to induce a certain reconciliation between these different hypotheses which are, however, opposed to each other.

The learned ignorance teaches us that our knowledge is relative and limited. Science alone would therefore not reveal to us the ultimate nature of reality, the absolute truth. It is preferable, in my opinion, to speak of conformity between observation and the interpretation of phenomena using a conceptual framework. The higher the level of conformity, the more likely and credible the conceptual framework used would be.

Metaphysical Reflections

Now, let's consider the following assumptions:

- The Universe is meaningful.
- The Universe is meaningless.

These hypotheses concern metaphysics, which can be defined as the quest for meaning. Metaphysics covers several fields of philosophy, including ontology, theology and cosmology. However, these fields are not disjointed from each other. They can be considered as different facets of the same representation of the world, provided that this representation is coherent.

I begin my reflection from the roots of ontology. Parmenides tells us that being exists and non-being (nothingness) does not exist. In a similar way, we can assume that sense is the attribute of being, and nonsense, the attribute of non-being. On the one hand, we know that nothingness has no meaning since it does not exist. On the other hand, it is logical to seek a meaning to what exists.

Some philosophers claim that existence itself is meaningless. This form of negative metaphysics seeks a meaningful explanation for what would be meaningless.

The hypothesis of nonsense is, in my opinion, a paradox comparable to that of non-being. It is an absurdity because it implies that the resulting reflection would itself be devoid of meaning. I conclude that, like non-being, nonsense does not exist.

If the Universe has a meaning, it is natural to seek to know what that meaning consists of, and where it comes from. This is where theology comes into play. For believers, only an all-knowing and all-powerful God can give meaning to the Universe he has himself engendered, a meaning that sometimes escapes human understanding. If this were not the case, the Universe would have no meaning and therefore would not exist.

In recent decades we have witnessed the rise of scientism, an atheistic ideology that postulates that the Universe created itself and that it is limited to what science can say about it. For my part, I believe that atheism leaves several fundamental questions unanswered:

- How could being emerge from non-being ?
- How could sense emerge from the nonsense ?
- How could complexity and life emerge from a global and increasing disorder?

- How could consciousness emerge from unconsciousness ?
- How could free will emerge from determinism ?

These reflections aim to show that the hypothesis of a transcendent and personal God is not only possible, but also provides a simple and coherent explanation for several existential problems. This belief, which dates to the dawn of time, remains to this day the one that rallies the greatest number of followers.

The Cusanian Theology

In the Learned Ignorance, Nicholas of Cusa compares God to the maximum sphere. From an ontological point of view, nothing can exist apart from the being himself. Since the physical universe cannot be external to God, God would have chosen to envelop it with his supreme being to maintain it in its bosom and monitor its evolution. Physically, this cosmic container is forever inaccessible, and yet since antiquity, it has been perfectly accessible to the understanding.

The Cardinal is obviously Christian and shows his deep attachment to the Holy Trinity. For him, the Absolute Maximum is simply irreducible. When Jesus prayed and performed miracles, he often lifted up his eyes to the sky, and when asked how to pray, he says explicitly: «Our Father, Who art in Heaven...» The association of the Christian God with the Cosmosphere is therefore quite natural. The Cosmosphere is in a way the Father's house, the heavenly dwelling place.

This cusanian theology is called Panentheism. Cusanian panentheism postulates the existence of a personal and transcendent God who maintains a particularly fruitful relationship with the Universe which he himself generated. It is necessary to distinguish cusanian panentheism from pantheism that can be associated with a form of naturalism.

The theological implications of The Learned Ignorance are beyond the scope of this book. Nicholas of Cusa is first a mystic who continues the work of Meister Eckhart and other theologians, whose thought is part of a form of Christian Neoplatonism called Rhenish mysticism. The interested reader is invited to refer to the cusanian works and the analyses made by specialists. In particular, the latest metaphysical work by Nicholas of Cusa entitled *«De non aliud»* offers us the image of a god both transcendent and immanent to the world (Cues, 2002). As the Absolute Maximum, God is by nature beyond opposites.

The Cosmosphere From a Theological Point of View

The cosmospheric system seeks to concretize this panentheistic theology through the Cosmosphere. This would be the means used by God to generate the physical universe and make it evolve. For a believer, the divine intervention is not limited to forging a universe in a hot big bang, and then abandoning it to its fate according to immutable parameters (deist position). By his omnipotence, God remains constantly present for his creation and can intervene at will in any field whatsoever. Indeed, how could we hope our prayers to be answered if they are addressed to a god who cannot act?

What we call chance would be a tool of divine intervention. Some claim that, because we understand more and more certain mechanisms of divine intervention, God is no longer required and therefore does not exist. From a logical point of view, this reasoning is not valid. For example, just because natural selection partly explains the mechanism of evolution does not mean that it is not part of the divine plan. It is a belief, not a proof.

In particular, the human being would be a privileged actor for the realization of the divine plan. Like all species, the human being belongs to nature and nature belongs to God. It is this filiation of man that makes us children of God. To exercise our free will, nature has endowed us with a mechanism to anticipate the future and guide it through our thoughts, words and actions. God has thus chosen to entrust us with a part of the work, talents to be made fruitful in this world still in the making. We are all, in one way or another, invited to contribute to the elaboration of a better world.

I believe that cosmology concerns primarily metaphysic and can hardly be confined to the scientific domain. I therefore reject the Big Bang theory in favor of a holistic cosmology, more coherent in my view.

In conclusion, the cosmospheric system described in this book is compatible with the panentheism from which it originates.

The Holographic Principle and the Omniscience of God

The holographic principle is a conjecture of the theory of quantum gravity proposed by Gerard't Hooft (Hooft, 1993), then improved by Leonard Susskind (Susskind, 1995). In essence, this principle postulates that the maximum quantity of information contained in a volume of space is limited by that stored on the surface of this volume. This quantity of information would be effectively maximal in the case of a black hole whose entropy is proportional to the surface of its horizon. This principle is now recognized by a growing community of physicists.

Since the Cosmosphere is the horizon of the cosmic black hole, it would be the Cosmosphere that constitutes the essence of reality. Our subjective perception of the physical world would be only the three-dimensional projection of this cosmic horizon, a little like a hologram.

If one associates the holographic principle with the cosmospheric principle, then the Cosmosphere contains all the information of the Universe. Whoever resides there has therefore the knowledge of the whole. Omniscience being a divine attribute, the holographic principle supports the cusanian hypothesis which assimilates the maximum sphere to God.

CONCLUSION

The concept of Cosmosphere dates to antiquity and the cosmospheric system embodies a conjecture dating from the Late Middle Ages. The cosmospheric hypothesis would have been prematurely discarded because of a widespread belief: objective realism.

For about a century now, the observation of quantum, relativistic and cosmological phenomena has brought to light a form of subjective realism. The Copernican principle on which the current theories are based thus derives from an arbitrary philosophical position. The same applies to the notions of space-time and the uniform flow of time. In short, the Big Bang theory is based on unjustified assumptions. For this reason, systemic cosmology is a valid and legitimate alternative to objective cosmology.

The cosmospheric system is the result of a structured process based on the systemic approach. It helps explain several phenomena, including inertia, gravitation, spatial expansion and its acceleration.

It avoids using many speculative and rather strange concepts, including physical infinities, the Big Bang, cosmic inflation, superluminal velocities, exotic dark matter, dark energy, extra dimensions, multiple universes and even space-time.

A new theory often raises more questions than it provides answers, and the cosmospheric system is no exception to this rule. This book does not pretend to identify all these questions, and much less answer them. However, because of its coherence, explanatory power and conformity to observation, I believe that the cosmospheric hypothesis deserves more attention.

If the cosmospheric hypothesis explained only two or three observations, one could consider them as simple coincidences. However, the abundance of compatible observations makes these coincidences highly improbable and the cosmospheric hypothesis all the more likely.

If this book had sowed doubt in your mind, then it was worth publishing. If some elements of its content seem compelling to you, then further analysis of those elements would be helpful, to say the least.

In your opinion, is the Universe really devoid of meaning and purpose? Would it have created itself, come out of nowhere by chance? Would it have generated living and conscious beings, always by chance? Would our very existence be meaningless, and is the free will we think we exercise just an illusion?

Whatever your opinion on the subject, I hope I have conveyed to you the taste to question the extent of our knowledge, as The Learned Ignorance of Nicholas of Cusa teaches us.

APPENDIX A - VALUES OF PHYSICAL CONSTANTS USED

Constant	Notation	Dimension	Value
Speed of light	c	$L\,T^{-1}$	299 792 458 $m\,s^{-1}$
Planck constant	h	$M\,L^{2}\,T^{-1}$	6.626 070 15 E-34 $kg\,m^{2}\,s^{-1}$
Reduced Planck constant	$\hbar$	$M\,L^{2}\,T^{-1}$	1.054 571 82 E-34 $kg\,m^{2}\,s^{-1}$
Gravitational constant	G	$M^{-1}\,L^{3}\,T^{-2}$	6.674 3 E-11 $kg^{-1}\,m^{3}\,s^{-2}$
Hubble constant *	H	T^{-1}	2.333 8 E-18 s^{-1}
Boltzman constant	k_{B}	$M\,L^{2}\,T^{-2}\,\theta^{-1}$	1.380 649 E-23 $J\,K^{-1}$
Coulomb Constant	k_{e}	$M\,L^{3}\,T^{-2}\,Q^{-2}$	8.987 551 792 E+09 $N\,m^{2}\,C^{-2}$
Elementary Charge	e	Q	1.602 176 634 E-19 C
Electron mass	m_{e}	M	9.109 383 701 5 E-31 kg
Proton mass	m_{p}	M	1.672 621 923 69 E-27 kg

CODATA 2018 values

* $H = G\,(m_e\,m_p)^{3/2}\,c^2/h\,k_e\,e^2$

APPENDIX B - MINIMUM VALUES OF PHYSICAL QUANTITIES

Quantity	Dimension	Minimum value	
		Formula	Value
Mass	M	$M_{min} = E_{min}/c^2$	1.720 6 E-68 kg
Length	L	$L_{min} = 2G\, M_{min}/c^2$	2.555 5 E-95 m
Time	T	$T_{min} = L_{min}/c$	8.524 1 E-104 s
Frequence	T^{-1}	$f_{min} = H$	2.333 8 E-18 s^{-1}
Velocity	$L\,T^{-1}$	$v_{min} = H\,L_{min}$	5.963 9 E-113 $m\,s^{-1}$
Acceleration	$L\,T^{-2}$	$a_{min} = H\,c$	6.996 5 E-10 $m\,s^{-2}$
Force	$M\,L\,T^{-2}$	$F_{min} = M_{min}\,a_{min}$	1.203 8 E-77 N
Energy	$M\,L^2\,T^{-2}$	$E_{min} = h\,H$	1.546 4 E-51 J
Action	$M\,L^2\,T^{-1}$	$A_{min} = h$	6.626 070 15 E-34 J s
Temperature	θ	$\theta_{min} = E_{min}/k_B$	1.120 0 E-28 K
Surface	L^2	$s_{min} = \pi\,L_{min}^2$	2.051 6 E-189 m^2
Volume	L^3	$V_{min} = (4/3)\pi\,L_{min}^3$	6.990 2 E-284 m^3

The accuracy of these values depends on the accuracy of the physical constants.

APPENDIX C - MAXIMUM VALUES OF PHYSICAL QUANTITIES

Quantity	Dimension	Maximum value	
		Formula	Value
Mass	M	$M_{max} = M_U = E_{max}/c^2$	8.649 0 E+52 kg
Length	L	$L_{max} = R_U = c/H$	1.284 6 E+26 m
Time	T	$T_{max} = 1/H$	4.284 9 E+17 s
Frequence	T^{-1}	$f_{max} = 1/T_{min}$	1.173 1 E+103 s^{-1}
Velocity	$L\ T^{-1}$	$v_{max} = c$	299 792 458 $m\ s^{-1}$
Acceleration	$L\ T^{-2}$	$a_{max} = f_{max}\ c$	3.517 0 E+111 $m\ s^{-2}$
Force	$M\ L\ T^{-2}$	$F_{max} = M_{max}\ a_{max}$	3.041 9 E+164 N
Energy	$M\ L^2\ T^{-2}$	$E_{max} = h\ f_{max}$	7.773 4 E+69 J
Action	$M\ L^2\ T^{-1}$	$A_{max} = E_{max}\ T_{max}$	3.330 8 E+87 J s
Temperature	θ	$\theta_{max} = E_{max}/k_B$	5.630 2 E+92 K
Surface	L^2	$s_{max} = \pi L_{max}^2$	5.184 1 E+52 m^2
Volume	L^3	$V_{max} = (4/3)\pi L_{max}^3$	8.879 2 E+78 m^3

The accuracy of these values depends on the accuracy of the physical constants.

APPENDIX D - INTERMEDIATE VALUES OF PHYSICAL QUANTITIES

Quantity	Dimension	Intermediate value	
		Formula	Value
Mass	M	$M_{int} = \sqrt{(M_{min} M_{max})}$	3.857 6 E-08 kg
Length	L	$L_{int} = \sqrt{(L_{min} L_{max})}$	5.729 5 E-35 m
Time	T	$T_{int} = \sqrt{(T_{min} T_{max})}$	1.911 1 E-43 s
Frequence	T^{-1}	$f_{int} = \sqrt{(f_{min} f_{max})}$	5.232 5 E+42 s^{-1}
Velocity	$L\,T^{-1}$	$v_{int} = \sqrt{(v_{min} v_{max})}$	1.337 1 E-52 $m\,s^{-1}$
Acceleration	$L\,T^{-2}$	$a_{int} = \sqrt{(a_{min} a_{max})}$	1.568 7 E+51 $m\,s^{-2}$
Force	$M\,L\,T^{-2}$	$F_{int} = \sqrt{(F_{min} F_{max})}$	6.051 3 E+43 N
Energy	$M\,L^2\,T^{-2}$	$E_{int} = \sqrt{(E_{min} E_{max})}$	3.467 1 E+09 J
Action	$M\,L^2\,T^{-1}$	$A_{int} = \sqrt{(A_{min} A_{max})}$	1.485 6 E+27 J s
Temperature	θ	$\theta_{int} = \sqrt{(\theta_{min} \theta_{max})}$	2.511 2 E+32 K
Surface	L^2	$s_{int} = \sqrt{(s_{min} s_{max})}$	1.031 3 E-68 m^2
Volume	L^3	$V_{int} = \sqrt{(V_{min} V_{max})}$	7.878 3 E-103 m^3

The accuracy of these values depends on the accuracy of the physical constants.

APPENDIX E - VALUES OF OTHER QUANTITIES USED

Quantity	Notation	Dimension	Value	Formula
Mass of the earth	M_T	M	5.972 2 E+24 kg	
Mass of Jupiter	M_J	M	1.898 6 E+27 kg	
Mass of the sun	$M_{\odot}$	M	1.988 4 E+30 kg	
Age of the Universe	H^{-1}	T	1.357 8 E+10 a	H^{-1}
Volume of the Universe	V_U	L^3	8.879 2 E+78 m^3	$4\pi c^3/3H^3$
Cosmic flow rate	$\dot{V}_U$	$L^3 T^{-1}$	6.216 6 E+61 $m^3 s^{-1}$	$4\pi c^3/H^2$
Cosmic flow rate variation	$\ddot{V}_U$	$L^3 T^{-2}$	1.450 8 E+44 $m^3 s^{-2}$	$4\pi c^3/H$
Volumic mass of the Universe	M_U / V_U	$M L^{-3}$	9.740 8 E-27 kg m^{-3}	$3H^2/8\pi G$
Critical Density of the Universe	$\rho_C = \rho_U$	$M L^{-1} T^{-2}$	8.754 6 E-10 J m^{-3}	$3c^2 H^2/8\pi G$
Entropy of the Universe	S_U	$M L^2 T^{-2} \theta^{-1}$	6.940 3 E+97 J K^{-1}	$k_B c^5/2G h H^2$
Cosmological constant	Λ	L^{-2}	1.818 0 E-52 m^{-2}	$3H^2/c^2$
Fine structure constant	α		7.297 353 E-03	$k_e e^2/\hbar c$
Force ratio e.m./grav.	β		2.268 7 E+39	$k_e e^2/G m_e m_p$
Max/min ratio	Σ		5.026 8 E+120	$c^5/2G h H^2$

The accuracy of these values depends on the accuracy of the physical constants.

APPENDIX F - VALUES OF A FICTITIOUS QUANTIZED QUANTITY (*Q*)

Index	Value
1	0.100 000
2	0.198 020
3	0.292 233
4	0.381 096
5	0.463 435
6	0.538 480
7	0.605 856
8	0.665 534
9	0.717 764
10	0.762 999
11	0.801 820
12	0.834 878
13	0.862 841
14	0.886 362
15	0.906 053
16	0.922 472
17	0.936 118
18	0.947 428
19	0.956 780
20	0.964 498
21	0.970 859
22	0.976 094
23	0.980 398
24	0.983 933
25	0.986 835

The quantum ($q = 0.1$) of this quantity Q would be excessively large compared to the maximum ($Q_{max} = 1$): $Q_1 = q$ and $Q_n = Q_{n-1} \tilde{+} q$.

APPENDIX G - THE *MELT-G* ENIGMA

The *MELT-hc* network in figure 5 shows known equations that express the wave nature of energy and mass. However, one can also formulate equations that involve the gravitational constant (G).

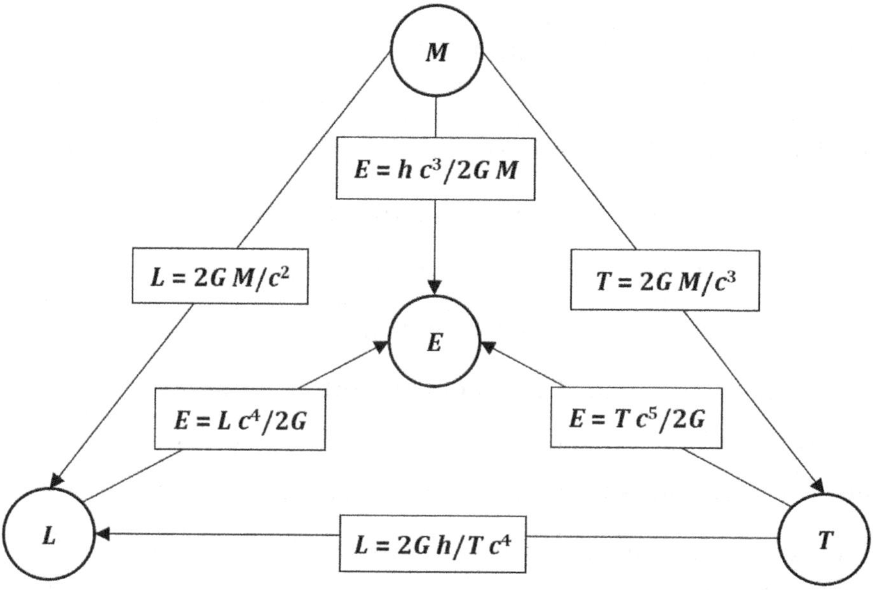

Figure 13 – The MELT-G Network

Apart from the Schwarzschild radius ($L = 2G\,M/c^2$) and its temporal link ($T = 2G\,M/c^3$), I don't know the physical meaning of these equations which seem to me yet algebraically correct. What is this energy which increases with time and decreases with mass? What about this length, inversely proportional to time?

Can you solve this enigma?

LIST OF FIGURES

TABLE OF CONTENTS

BIBLIOGRAPHY

Barnes, L., & Lewis, G. (2020). *The Cosmic Revolutionary's Handbook.* Cambridge: Cambridge University Press.

Barrau, A. (2016). *De la vérité dans les sciences.* Paris: Dunod.

Barrow, J. D. (1996). *La grande théorie.* Paris: Flammarion.

Barrow, J. D. (2005). *The infinite book.* London: Vintage.

Bouazaoui, M., Delahaye, J.-P., & Wlodarczak, G. (2003). *L'infini dans les sciences, l'art et la philosophie.* Paris: L'Harmattan.

Broglie, L. d. (1924). *Recherches sur la théorie des Quanta.* Annales de la fondation Louis de Broglie. France: Migration - Université en cours d'affectation.

Carvalho, J. (1995). Derivation of the mass of the observable universe. *International Journal of Theoretical Physics.*

Cazenave, M. (1996). *La science et l'âme du monde.* Paris: Albin Michel.

Cazenave, M. (2005). *De la science à la philosophie.* France: Albin Michel.

Chardin, P. T. (1955). *Le phénomène humain.* France: Le Seuil.

Cohen-Tannoudji, G. (1998). *Les constantes universelles.* France: Hachette.

Crescenzo, L. d. (1999). *Les grands philosophes de la Grèce antique.* Milan: Éditions de Fallois.

Cues, N. d. (2002). *Du non-autre - Le guide du penseur.* Paris: Les éditions du cerf.

Cues, N. d. (2008). *La Docte Ignorance.* (H. Pasqua, Trans.) Paris: Éditions Payot & Rivages.

Davis, P. (1983). *God and the new physics.* New York: Simon & Schuster.

Davis, P. (1993). *The mind of God.* New York: Simon & Schuster.

Dirac, P. (1937). The Cosmological Constants. *Nature.*

During, É. (1998). *La métaphysique.* Paris: Flammarion.

Einstein, A. (1956). *La relativité.* Saint-Amand (Cher.): Éditions Gauthier-Villars.

Einstein, A., & Infeld, L. (1983). *L'évolution des idées en physique.* Paris: Flammarion.

Fahr, H., & Heyl, M. (2007). Cosmic vacuum energy decay and creation of cosmic matter. *Die Naturwissenschaften.*

Farouki, N. (1995). *La métaphysique.* France: Flammarion.

Gauthier, Y. (1995). *La philosophie des sciences.* Montréal: Les presses de l'Université de Montréal.

Grondin, J. (2004). *Introduction à la métaphysique.* Montréal: Presses de l'Université de Montréal.

Haranas, I., & Harney, M. (2009). The Mass of the Universe and Other Relations in the Idea of a Possible Cosmic Quantum Mechanics. *Progress in Physics.*

Hawking, S., & Penrose, R. (1997). *La nature de l'espace et du temps.* France: Gallimard.

Hooft, G. '. (1993). Dimensional Reduction in Quantum Gravity. *General Relativity and Quantum Cosmology,* 13.

Hoyle, F. (1948). A new model for the expanding universe. *Monthly Notices of the Royal Astronomical Society.*

Hoyle, F. (1958). The Steady State Theory. *La structure et l'évolution de l'Univers.* Bruxelles: Institut international de physique Solvay.

Hudry, F. (1989). *Le livre des XXIV philosophes.* Paris: Librairie philosophique J. Vrin.

Iorio, L. (2011). An empirical explanation of the anomalous increases in the astronomical unit and the lunar eccentricity. *The Astronomical Journal*, 3.

Iorio, L. (2011). On the anomalous secular increase of the eccentricity of the orbit of the Moon. *Monthly Notices of the Royal Astronomical Society*, 10.

Jeans, J. (1928). The Physics of the Universe. *Nature.*

Kant, E. (1944). *Critique de la raison pure.* Paris: Presses universitaires de France.

Klein, É., & Lachièze-Rey, M. (1996). *La quête de l'unité.* France: Albin Michel.

Klein, É., & Spiro, M. (1996). *Le temps et sa flèche.* France: Flammarion.

Koyré, A. (1973). *Du monde clos à l'Univers infini.* France: Éditions Galimard.

Languirand, J., & Proulx, J. (2008). *Le Dieu cosmique.* Montréal: Le Jour éditeur.

Lepeltier, T. (2014). *La face cachée de l'Univers - Une autre histoire de la cosmologie.* Paris: Éditions du Seuil.

Lepeltier, T. (2021). *L'Univers existe-t-il ?* Paris: Presses Universitaires de France.

Lerner, E. J. (1992). *The Big Bang Never Happened.* New York: Vintage Books.

Lloyd, S. (2002, May 24). Computational Capacity of the Universe. *Physical review letters*, p. 4.

Marty, B., Zimmermann, L., Pujol, M., Burgess, R., & Philippot, P. (2013). Nitrogen Isotopic Composition and Density of the Archean Atmosphere. *Science, American Association for the Advancement of Science.*

Milgrom, M. (2020). MOND vs. dark matter in light of historical parallels. *Astrophysics of Galaxies (astro-ph.GA).*

Monnoyeur, F. (1992). *Infini des mathématiciens, infini des philosophes.* France: Bélin.

Monnoyeur, F. (1995). *Infini des philosophes, infini des astronomes.* France: Bélin.

Narlikar, J. (1993). *Introduction to Cosmology.* Cambridge: Cambridge University Press.

Narlikar, J., & Arp, H. (1993). Flat Spacetime Cosmology: A Unified framework for extragalactic redshifts. *The Astrophysical Journal*, 6.

Narlikar, J., & Arp, H. (1997). Time Dilation in the Supernova Light Curve and the Variable Mass Hypothesis. *THE ASTROPHYSICAL JOURNAL.*

Nicolle, J.-M. (n.d.). *La coïncidence des opposés.* Retrieved from Nicolas de Cues (1401-1464): http://jm.nicolle.cusa.pagesperso-orange.fr/glossaire/coincidence.htm

Penrose, R. (1999). *Les deux infinis et l'esprit humain.* France: Flammarion.

Penrose, R. (2004). *The Road to Reality.* London: Jonathan Cape.

Penrose, R. (2007). *À la découverte des lois de l'Univers.* Paris: Odile Jacob.

Peter, P., & Uzan, J.-P. (2005). *Cosmologie primordiale.* France: Bélin.

Proulx, J. (2001, printemps). Retrouver la dimension cosmologique de la théologie chrétienne. *Théologiques, 9, (1), 49–70. https://doi.org/10.7202/005689ar*, pp. 49-70.

Proulx, J. (2009). *La chorégraphie divine - Essai sur le cosmos.* St-Laurent, Qc: Septentrion.

Proulx, J. (2011). *En quête de sens.* Montréal: Le Jour éditeur.

Sagan, C., & Mullen, G. (1972). Earth and Mars: Evolution of Atmospheres and Surface Temperatures. *Science.*

Sanjoy M. Som, R. B. (2016). Earth's air pressure 2.7 billion years ago constrained to less than half of modern levels. *Nature Geoscience.*

Séguin, M., & Villeneuve, B. (1995). *Astronomie et astrophysique.* Saint-Laurent, Qc: Éditions du Rrenouveau Pédagogique.

Seife, C. (2007). *Decoding the Universe.* United States: Pengouin books.

Smolin, L. (2007). *Rien ne va plus en physique !* Paris: Dunod.

Susskind, L. (1995). The World as a Hologram. *Journal of Mathematical Physics.*

Talbot, M. (2011). *The Holographic Universe.* New York: Harper Collins.

Talbot, P. (2021). *Le principe cosmosphérique.* (P. Talbot, Ed.) St-Hubert, Qc, Québec, Canada: Paul Talbot.

Thom, R., Castoriadis, C., Demaret, J., Lambert, D., Rocher, G., & Lane, G. (1994). *La Magie contemporaine - L'échec du savoir moderne.* Cap-St-Ignace, Qc: Québec/Amérique.

Valev, D. (2013). Three Fundamental Masses Derived by Dimensional Analysis. *American Journal of Space Science* (p. 5). Science Publications.

Wesson, P. S. (2003). *Is mass quantized.* Waterloo, CA: University of Waterloo.

Wilhelm, K., & Dwivedi, B. (2011). An Explanation of the Pioneer anomaly involving accelerated atomic clocks. *Astrophysics and Space Sciences Transactions*, 8.

Wilhelm, K., & Dwividi, B. (2013). Increase of the mean Sun–Earth distance caused by a secular mass accumulation. *Astrophys Space Sci.*

Made in the USA
Monee, IL
04 November 2026

67379315R00105